T0155570

Tag Counting and Monitoring in Large-Scale
RFID Systems

Jihong Yu • Lin Chen

Tag Counting and Monitoring in Large-Scale RFID Systems

Theoretical Foundations and Algorithm Design

 Springer

Jihong Yu
Simon Fraser University
Burnaby, BC, Canada

Lin Chen
Laboratoire de Recherche en Informatique
University of Paris-Sud
Orsay, France

ISBN 978-3-030-06344-3   ISBN 978-3-319-91992-8   (eBook)
https://doi.org/10.1007/978-3-319-91992-8

Printed on acid-free paper

This Springer imprint is published by the registered company Springer International Publishing AG part
of Springer Nature.
The registered company address is: Gewerbestrasse 11, 6330 Cham, Switzerland

Preface

Radio frequency identification (RFID) technology has been experiencing ever-increasing deployment in a wide range of various applications, such as inventory control and supply chain management. In this book, we present systematic research on a number of research problems related to tag counting and monitoring, one of the most fundamental components in RFID systems, particularly when the system scales. These problems are simple to state and intuitively understandable, while of both fundamental and practical importance, and require nontrivial efforts to solve. Specifically, we address the following problems ranging from theoretical modelling and analysis to practical algorithm design and optimisation:

- Stability analysis of the Frame Slotted Aloha (FSA) protocol, the *de facto* standard in RFID tag counting and identification,
- Tag population estimation in dynamic RFID systems,
- Missing tag event detection in the presence of unexpected tags,
- Missing tag event detection in multiple-group multiple-region RFID systems.

In the book, we adopt a research and exposition line from theoretical modelling and analysis to practical algorithm design and optimisation.

To lay the theoretical foundations for the design and optimisation of tag counting and monitoring algorithms, we start by investigating the stability of FSA. Technically, we model the system backlog as a Markov chain with its states being backlog size at the beginning of each frame. The main objective is mathematically translated to analyse the ergodicity of the Markov chain and derive its properties in different regions including the instability region. By employing drift analysis, we derive the closed-form conditions for the stability of FSA and find the stability region maximiser. We also mathematically demonstrate the existence of transience of the backlog Markov chain, which characterises system behaviour in the instability region.

We then establish a generic framework of stable and accurate tag population estimation schemes based on Kalman filter for both static and dynamic RFID systems. Specifically, we model the dynamics of RFID systems as discrete stochastic processes and leverage the techniques in extended Kalman filter (EKF) and cumulative sum control chart (CUSUM) to estimate tag population for both static and dynamic systems. By employing Lyapunov drift analysis, we mathematically characterise the performance of our framework in terms of estimation accuracy and convergence speed by deriving the closed-form conditions on the design parameters under which our scheme can stabilise around the real population size with bounded relative estimation error that tends to zero with exponential convergence rate.

We further proceed to addressing the problem of missing tag detection, one of the most important RFID applications. Different to existing works in this field, we focus on two unexplored while fundamentally important scenarios, missing tag detection in the presence of unexpected tags and in multiple-group multiple-region RFID systems. In the first scenario, we develop a two-phase Bloom filter-based missing tag detection protocol (BMTD). The proposed BMTD exploits Bloom filter in sequence to first deactivate the unexpected tags and then test the membership of the expected tags, thus dampening the interference from the unexpected tags and considerably reducing the detection time. To minimise the detection time of BMTD while achieving the required reliability, we perform theoretical analysis and optimisation on configuring the protocol parameters. In the second scenario, we formulate and study a new missing tag detection problem, arising in multiple-group multiple-region RFID systems, where a mobile reader needs to detect whether there is any missing event for each group of tags. The objective is to devise missing tag detection protocols with minimum execution time while meeting the detection reliability requirement for each group. We develop a suite of three missing tag detection protocols, each decreasing the execution time compared to its predecessor by incorporating an improved version of the Bloom filter design and parameter tuning. By sequentially analysing the developed protocols, we gradually iron out an optimum detection protocol that works in practice.

Burnaby, BC, Canada   Jihong Yu
Orsay, France   Lin Chen

Contents

Chapter 1
Introduction

1.1 RFID Technology

Radio frequency identification (RFID) is one of the most promising and widely used automatic identification and data capture technologies. RFID technology is the use of radio waves to read, capture, and interact with simple, lightweight and attachable tags that store information about physical objects. Therefore, with a tag attached on a physical object RFID technology is able to monitor this object uniquely in real time. Moreover, RFID technology supports non-line-of-sight communication and has longer communication range, which overcomes the drawbacks of traditional barcode technology [1]. Recent years have witnessed an unprecedented development of the RFID technology for its low cost and its potentiality enabling Internet of Things [2]. The RFID market was worth over $11 billion, and forecast to reach about $15 billion in 5 years [3].

An RFID system typically consists of one or several RFID readers and a large number of RFID tags. Specifically, an RFID reader is a device equipped with a dedicated power source and an antenna. The reader can wireless collect and process the information of tags within its coverage area. An RFID tag, on the other hand, is a low-cost microchip labeled with a unique serial number (ID) to identify a physical object and can receive/transmit the radio signals via the wireless channel. More specifically, the tags are generally classified into three categories: passive tags, active tags, and semi-active tags. A passive tag does not need a battery, in contrast, it captures energy in the radio frequency signal from its nearby reader for normal operations, and backscatters its messages to the reader. Whereas, an active tag has on-board battery and periodically transmits its signal. Compared with passive and active tags, a semi-active tag has a smaller battery on board but is activated just when in the presence of a reader. To facilitate the communication between readers and tags, EPCGlobal C1G2 [4], one of the most popular industrial standards for RFID, specifies frame slotted Aloha (FSA) as the medium access control protocol.

J. Yu, L. Chen, *Tag Counting and Monitoring in Large-Scale RFID Systems*,
https://doi.org/10.1007/978-3-319-91992-8_1

RFID systems have been deployed in real-world scenarios for various applications ranging from inventory control [5, 6] and supply chain management [7] to object tracking [8] and localization [9]. In most, if not all, RFID applications, tag counting and monitoring are perhaps one of the most fundamental component. Although simple to state and intuitively understandable, designing efficient tag counting and monitoring algorithms require non-trivial efforts to solve, especially in large-scale RFID systems, due to the following particular challenges in RFID systems.

- *Large number of tags*. An RFID system may consist of a large number of tags, such as a warehouse storing thousands of goods for retailers. Any effective algorithm designed for these RFID systems needs to scale elegantly.
- *Limited computing resource at tags*. The quest of compatible size and low energy consumption significantly limits the computing and processing capability of individual tags in RFID systems, especially for lightweight passive tags.
- *Unreliable wireless links*. Wireless links are notoriously unreliable and error-prone. Hence, algorithms should be robust in the sense that they are able to work under unreliable channel conditions.

The challenges bring out new issues on efficient tag counting and monitoring in large-scale RFID systems. This book systematically presents state-of-the-art tag counting and monitoring protocols, which focuses on several representative research problems of both fundamental and practical importance. Specifically, we address the following problems ranging from theoretical modeling and analysis, to practical algorithm design and optimisation.

- Stability analysis of the frame slotted Aloha (FSA) protocol, the *de facto* standard in RFID tag counting and identification,
- Tag population estimation in dynamic RFID systems,
- Missing tag event detection in the presence of unexpected tags,
- Missing tag event detection in multiple-group multiple-region RFID systems.

1.2   FSA Stability

To lay the theoretical foundations for the design and optimization of tag counting and monitoring algorithms, we start by investigating the stability of FSA, which is of fundamental importance both on the theoretical characterisation of FSA performance and its effective operation in practical systems. To study the stability of FSA, the effort should be devoted to answering the following natural and crucial questions:

- Under what condition(s) is FSA stable?
- When is the stability region maximised?
- How does FSA behave in the instability region?

While the above fundamental questions have not been explored in the literature. To fill the void in the study of FSA stability, we investigate these questions by technically modeling the FSA system backlog as a Markov chain with its states being backlog size at the beginning of each frame. The main objective is to analyze the ergodicity of the Markov chain and demonstrate its properties in different regions, particularly the instability region. Specifically, by employing drift analysis, we obtain the closed-form conditions for the stability of FSA and show that the stability region is maximised when the frame length equals the number of packets to be sent in the single packet reception model and the upper bound of stability region is maximised when the ratio of the number of packets to be sent to frame length equals in an order of magnitude the maximum multipacket reception capacity in the multipacket reception model, which answers the first two questions. Furthermore, to characterise system behavior in the instability region, we mathematically demonstrate the existence of transience of the backlog Markov chain, which provides the answer to the third question.

1.3   Tag Counting

The problem of tag counting, or tag population estimation, which is to fast and accurately estimate the number of tags in the current interrogation region of RFID readers, has recently attracted significant research attention due to its paramount importance on a variety of RFID applications. However, most, if not all, of existing estimation mechanisms [10–14] are proposed for the static case where tag population remains constant during the estimation process, thus leaving the more challenging dynamic case unaddressed, despite the fundamental importance of the latter case on both theoretical analysis and practical application.

Motivated by the above argument, we design a generic framework of stable and accurate tag population estimation schemes based on Kalman filter for both static and dynamic RFID systems. Our main contributions are twofold. Firstly, we model the dynamics of RFID systems as discrete stochastic processes and leverage the techniques in extended Kalman filter (EKF) and cumulative sum control chart (CUSUM) to estimate tag population for both static and dynamic systems. Secondly, By employing Lyapunov drift analysis, we mathematically characterise the performance of the proposed framework in terms of estimation accuracy and convergence speed by deriving the closed-form conditions on the design parameters under which our scheme can stabilise around the real population size with bounded relative estimation error that tends to zero with exponential convergence rate.

1.4   Tag Monitoring

RFID technology has been widely used in missing tag detection to reduce and avoid inventory shrinkage by monitoring tags attaching on products. Obviously, the first step in the application of loss prevention is to determine whether there is any missing tag, which is beneficial to find theft events in warehouses or stores and reduce financial loss. Promptly finding out the missing tag event is thus of practical importance. Different from existing works on missing tag detection [15–21], we focus on two unexplored while fundamentally important scenarios, missing tag detection in the presence of unexpected tags and in multiple-group multiple-region RFID systems.

Missing Tag Detection in RFID Systems with the Presence of Unexpected Tags
In the first scenario, existing missing tag detection protocols cannot efficiently handle the presence of a large number of unexpected tags whose IDs are not known to the reader, which shackles the time efficiency. To deal with the problem of detecting missing tags in the presence of unexpected tags, we devise a two-phase Bloom filter-based missing tag detection protocol (BMTD). The proposed BMTD exploits Bloom filter in sequence to first deactivate the unexpected tags and then test the membership of the expected tags, thus dampening the interference from the unexpected tags and considerably reducing the detection time. Moreover, the theoretical analysis of the protocol parameters is performed to minimize the detection time of the proposed BMTD and achieve the required reliability simultaneously. Extensive experiments are then conducted to evaluate the performance of the proposed BMTD. The results demonstrate that the proposed BMTD significantly outperforms the state-of-the-art solutions.

Missing Tag Detection in Multiple-Group Multiple-Region RFID Systems We formulate and study a new missing tag detection problem, arising in multiple-group multiple-region RFID systems, where a mobile reader needs to detect whether there is any missing event for each group of tags. The problem we tackle is to devise missing tag detection protocols with minimum execution time while guaranteeing the detection reliability requirement for each group. By leveraging the technique of Bloom filter, we develop a suite of three missing tag detection protocols, each decreasing the execution time compared to its predecessor by incorporating an improved version of the Bloom filter design and parameter tuning. By sequentially analysing the developed protocols, we gradually iron out an optimum detection protocol that works in practice.

1.5   Book Organization

In this book, we adopt a research and exposition line from theoretical modeling and analysis to practical algorithm design and optimisation. Figure 1.1 illustrates the structure of the book. In the remainder of this section, we provide a high-level

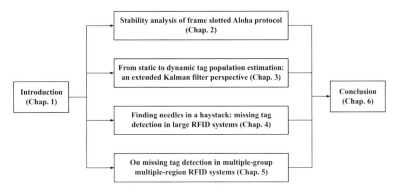

Fig. 1.1 Book organization

overview of the technical contributions of this book, which are presented sequentially in Chaps. 2–5. To facilitate readers, we adopt a modularized structure to present the results such that the chapters are arranged as independent modules, each devoted to a specific topic outlined above. In particular, each chapter has its own introduction and conclusion sections, elaborating the related work and the importance of the results with the specific context of that chapter. For this reason, we are not providing a detailed background, or a survey of prior work here.

References

1. Barcode [Online] (2016), https://en.wikipedia.org/wiki/Barcode
2. F. Mattern, C. Floerkemeier, From the internet of computers to the internet of things, in *From Active Data Management to Event-Based Systems and More* (Springer, Berlin, 2010), pp. 242–259
3. Rfid forecasts, players and opportunities 2017–2027 [Online] (2017), https://www.idtechex.com/research/reports/rfid-forecasts-players-and-opportunities-2017-2027-000546.asp
4. EPCglobal Inc., Radio-frequency identity protocols class-1 generation-2 UHF RFID protocol for communications at 860 MHz–960 MHz version 1.0.9, EPCglobal Inc., vol. 17 (2005)
5. RFID Journal, DoD releases final RFID policy. [Online]
6. RFID Journal, DoD reaffirms its RFID goals. [Online]
7. C.-H. Lee, C.-W. Chung, Efficient storage scheme and query processing for supply chain management using RFID, in *ACM SIGMOD* (ACM, New York, 2008), pp. 291–302
8. L.M. Ni, D. Zhang, M.R. Souryal, RFID-based localization and tracking technologies. IEEE Wirel. Commun. **18**(2), 45–51 (2011)
9. P. Yang, W. Wu, M. Moniri, C.C. Chibelushi, Efficient object localization using sparsely distributed passive RFID tags. IEEE Trans. Ind. Electron. **60**(12), 5914–5924 (2013)
10. M. Kodialam, T. Nandagopal, W.C. Lau, Anonymous tracking using RFID tags, in *IEEE INFOCOM* (IEEE, Piscataway, 2007), pp. 1217–1225
11. T. Li, S. Wu, S. Chen, M. Yang, Energy efficient algorithms for the RFID estimation problem, in *IEEE INFOCOM* (IEEE, Piscataway, 2010), pp. 1–9
12. C. Qian, H. Ngan, Y. Liu, L.M. Ni, Cardinality estimation for large-scale RFID systems. IEEE Trans. Parallel Distrib. Syst. **22**(9), 1441–1454 (2011)

13. M. Shahzad, A.X. Liu, Every bit counts: fast and scalable RFID estimation, in *ACM Mobicom* (2012), pp. 365–376
14. Y. Zheng, M. Li, Zoe: fast cardinality estimation for large-scale RFID systems, in *IEEE INFOCOM* (IEEE, Piscataway, 2013), pp. 908–916
15. C.C. Tan, B. Sheng, Q. Li, How to monitor for missing RFID tags, in *IEEE ICDCS* (IEEE, Piscataway, 2008), pp. 295–302
16. W. Luo, S. Chen, T. Li, Y. Qiao, Probabilistic missing-tag detection and energy-time tradeoff in large-scale RFID systems, in *ACM MobiHoc* (ACM, New York, 2012), pp. 95–104
17. W. Luo, S. Chen, Y. Qiao, T. Li, Missing-tag detection and energy–time tradeoff in large-scale RFID systems with unreliable channels. IEEE/ACM Trans. Netw. **22**(4), 1079–1091 (2014)
18. M. Shahzad, A.X. Liu, Expecting the unexpected: fast and reliable detection of missing RFID tags in the wild, in *IEEE INFOCOM* (2015), pp. 1939–1947
19. T. Li, S. Chen, Y. Ling, Identifying the missing tags in a large RFID system, in *ACM MobiHoc* (ACM, New York, 2010), pp. 1–10
20. R. Zhang, Y. Liu, Y. Zhang, J. Sun, Fast identification of the missing tags in a large RFID system, in *IEEE SECON* (IEEE, Piscataway, 2011), pp. 278–286
21. X. Liu, K. Li, G. Min, Y. Shen, A.X. Liu, W. Qu, Completely pinpointing the missing RFID tags in a time-efficient way. IEEE Trans. Comput. **64**(1), 87–96 (2015)

Chapter 2
Stability Analysis of Frame Slotted Aloha Protocol

Chapter Roadmap The rest of this chapter is organised as follows. Section 2.1 explains the motivation of studying FSA stability and summarizes the contributions. Section 2.2 gives a brief overview of related work and compares our results with existing results. In Sect. 2.3, we present the system model, including random access model, traffic model and packet success probability. In Sect. 2.4, we summary the main result of this chapter and the detailed proofs on the stability properties of FSA-SPR and FSA-MPR are given in Sects. 2.5 and 2.6, respectively. In Sect. 2.7, we study the frame size control in practice. In Sect. 2.8, we conduct the numerical analysis. Finally, we conclude the chapter in Sect. 2.9.

2.1 Introduction

2.1.1 Context and Motivation

Since the introduction of Aloha protocol in 1970 [1], a variety of such protocols have been proposed to improve its performance, such as Slotted Aloha (SA) [2] and Frame Slotted Aloha (FSA) [3]. SA is a well known random access scheme where the time of the channel is divided into identical slots of duration equal to the packet transmission time and the users contend to access the server with a predefined slot-access probability. As a variant of SA, FSA divides time-slots into *frames* and a user is allowed to transmit only a single packet per frame in a randomly chosen time-slot.

Due to their effectiveness to tackle collisions in wireless networks, SA- and FSA-based protocols have been applied extensively to various networked systems ranging from the satellite networks [4], wireless LANs [5, 6] to the emerging Machine-to-Machine (M2M) networks [7, 8]. Specifically, in radio frequency identification (RFID) systems, which is our specific interest in this book, FSA

© Springer International Publishing AG, part of Springer Nature 2019
J. Yu, L. Chen, *Tag Counting and Monitoring in Large-Scale RFID Systems*,
https://doi.org/10.1007/978-3-319-91992-8_2

plays a fundamental role in the identification of tags [9, 10] and is standardized in the EPCGlobal Class-1 Generation-2 (C1G2) RFID standard [11]. In FSA-based protocols, all users with packets transmit in the selected slot of the frame respectively, but only packets experiencing no collisions are successful while the other packets referred to as backlogged packets (or simply backlogs), are retransmitted in the subsequent frames.

Given the paramount importance of the stability for systems operating on top of Aloha-based protocols, a large body of studies have been devoted to stability analysis in a slotted collision channel [12–14] where a transmission is successful if and only if just a single user transmits in the selected slot, referred to as single packet reception (SPR). Differently with SPR, the emerging multipacket reception (MPR) technologies in wireless networks, such as Code Division Multiple Access (CDMA) and Multiple-Input and Multiple-Output (MIMO), make it possible to receive multiple packets in a time-slot simultaneously, which remarkably boosts system performance at the cost of the system complexity.

More recently, the application of FSA in RFID systems and M2M networks has received considerable research attention. However, very limited work has been done on the stability of FSA despite its fundamental importance both on the theoretical characterisation of FSA performance and its effective operation in practical systems. Motivated by the above observation, we argue that a systematic study on the stability properties of FSA incorporating the MPR capability is called for in order to lay the theoretical foundations for the design and optimization of FSA-based communication systems.

2.1.2 Summary of Contributions

In this chapter, we investigate the stability properties of p-persistent FSA with SPR and MPR capabilities. The main contributions of this chapter are articulated as follows:

- Firstly, we model the packet transmission process in a frame as the bins and balls problem [15] and derive the number of successfully received packets under both SPR and MPR models.
- Secondly, we formulate a homogeneous Markov chain to characterize the number of the backlogged packets and derive the one-step transition probability with the persistence probability p.
- Thirdly, by employing drift analysis, we obtain the closed-form conditions for the stability of p-persistent FSA and derive conditions maximising the stability regions for both SPR and MPR models.
- Fourthly, to characterise system behavior in the instability region with the persistence probability p, we mathematically demonstrate the existence of transience of the backlog Markov chain.

- Fifthly, we investigate how to achieve the stability condition and give the control algorithm for updating the frame size.
- Finally, we conduct extensive experiments to demonstrate the analytical results.

Our work demonstrates that the stability region is maximised when the frame length equals the number of sent packets in the SPR model and the upper bound of stability region is maximised when the ratio of the number of sent packets to frame length equals in an order of magnitude the maximum multipacket reception capacity in the MPR model. In addition, it is also shown that FSA-MPR outperforms FSA-SPR remarkably in terms of the stability region size.

2.2   Related Work

Aloha-based protocols are basic schemes for random medium access and are applied extensively in many communication systems. As a central property, the stability of Aloha protocols has received a lot of research attention, which we briefly review in this section.

Stability of Slotted Aloha Tsybakov and Mikhailov [16] initiated the stability analysis of finite-user slotted Aloha. They found sufficient conditions for stability of the queues in the system using the principle of stochastic dominance and derived the stability region for two users explicitly. For the case of more than two users, the inner bounds to the stability region were shown in [17]. Subsequently, Szpankowski [18] found necessary and sufficient conditions for the stability under a fixed transmission probability vector for three-user case. However, the derived conditions are not closed-form, meaning the difficulty on verifying them. In [12] an approximate stability region was derived for an arbitrary number of users based on the mean-field asymptotics. It was claimed that this approximate stability region is exact under large user population and it is accurate for small-sized networks. The sufficient condition for the stability was further derived to be linear in arrival rates without the requirement on the knowledge of the stationary joint statistics of queue lengths in [13]. Recently, the stability region of SA with K-exponential backoff was derived in [14] by modeling the network as inter-related quasi-birth-death processes. We would like to point out that all the above stability analysis results were derived for the SPR model.

Stability of Slotted Aloha with MPR The first attempt at analyzing stability properties of SA with MPR was made by Ghez et al. in [19, 20] in an infinite-user single-buffer model. They drew a conclusion that the system could be stabilized under the symmetrical MPR model with a non-zero probability that all packets were transmitted successfully. Afterwards, Sant and Sharma [21] studied a special case of the symmetrical MPR model for finite-user with an infinite buffer. They derived sufficient conditions on arrival rate for stability of the system under the stationary ergodic arrival process. Subsequently, the effect of MPR on stability and delay was investigated in [22] and it was shown that stability region undergoes a phase transition and then reaches the maximization. Besides, in [23] necessary and sufficient conditions are obtained for a Nash equilibrium strategy for wireless networks with MPR based on noncooperation game theory. More recently, Jeon and Ephremides [24] characterised the exact stability region of SA with stochastic energy harvesting and MPR for a pair of bursty users. Although the work aforementioned analyzed the stability of system without MPR or/and with MPR, they are mostly, if not all, focused on SA protocol, while our focus is FSA with both SPR and MPR.

Performance Analysis of FSA There exist several studies on the performance of FSA. Wieselthier and Anthony [25] introduced an combinatorial technique to analyse performance of FSA-MPR for the case of finite users. Schoute [26] investigated dynamic FSA and obtained the expected number of time-slots needed until the backlog becomes zero. Recently, the optimal frame setting for dynamic FSA was proved mathematically in [27] and [28]. However, these works did not address the stability of FSA, which is of fundamental importance.

In summary, only very limited work has been done on the stability of FSA despite its fundamental importance both on the theoretical characterisation of FSA performance and its effective operation in practical systems. In order to bridge this gap, we devote this chapter to investigating the stability properties of FSA under both SPR and MPR models.

2.3   System Model

In this section, we introduce our system model which will be used throughout the rest of this chapter.

2.3.1   Physical Layer and Random Access Model in FSA

We consider a system of infinite identical users operating on one frequency channel. In one slot, a node can complete a packet transmission. We investigate two physical layer models of practical importance, the models with single packet reception (SPR) and multipacket reception (MPR) capabilities:

- Under the SPR model, a packet suffers a collision if more than one packet is transmitted in the same time-slot. SPR is a classical and baseline physical layer model.
- Under the MPR model, up to $\overline{M}$ ($\overline{M} > 1$) concurrently transmitted packets can be received successfully with non-zero probabilities as specified by a stochastic matrix Ξ defined as follows:

$$\Xi \triangleq \begin{pmatrix} \hat{\xi}_{10} & \hat{\xi}_{11} \\ \hat{\xi}_{20} & \hat{\xi}_{21} & \hat{\xi}_{22} \\ \vdots & \vdots & \vdots & \ddots & & & 0 \\ \hat{\xi}_{x_00} & \hat{\xi}_{x_01} & \cdots\cdots & \cdots & \hat{\xi}_{x_0x_0} \\ \vdots & \vdots & \vdots & \vdots & & \ddots \\ \hat{\xi}_{\overline{M}0} & \hat{\xi}_{\overline{M}1} & \cdots\cdots & \cdots & \hat{\xi}_{\overline{M}\,\overline{M}} \\ 1 & 0 & \cdots\cdots & \cdots & 0 \end{pmatrix} \tag{2.1}$$

where $\hat{\xi}_{x_0k_0}$ ($k_0 \leq x_0 \leq \overline{M}$) is the probability of having k_0 successful packets among x_0 transmitted packets in one slot. Ξ is referred to as the reception matrix. The last two decades have witnessed an increasing prevalence of MPR technologies such as CDMA and MIMO. Mathematically, the SPR model can be regarded as a degenerated MPR model with $\overline{M} = 1$ and

$$\Xi = \begin{pmatrix} 0 & 1 \\ 1 & 0 \\ \vdots & \vdots & 0 \\ 1 & 0 \end{pmatrix}.$$

The random access process operates as follows: FSA organises time-slots with each frame containing a number of consecutive time-slots. Each user is allowed to randomly and independently choose a time-slot to send his packet at most once per frame. More specifically, suppose the length of frame t is equal to L_t, then in the beginning of frame t each user generates a random number $\mathfrak{R}$ and selects the $(\mathfrak{R} \mod L_t)$-th time-slot in frame t to transmit his packet. Note that unsuccessful packets in the current frame are retransmitted in the next frame with the constant persistence probability p while newly generated packets are transmitted in the next frame following their arrivals with probability one.

For notation convenience, we use FSA-SPR and FSA-MPR to denote the FSA system operating on the SPR and MPR models, respectively.

2.3.2   Traffic Model

Let random variable N_t denote the total number of new arrivals during frame t and denote by A_{tl} the number of new arrivals in time-slot l in frame t where $l = 1, 2, \cdots, L_t$. Assume that (A_{tl}) are independent and identically Poisson distributed random variables with probability distribution:

$$P\{A_{tl} = u\} = \Lambda_u (u \geq 0) \tag{2.2}$$

such that the expected number of arrivals per time-slot $\Lambda = \sum_1^\infty u\Lambda_u$ is finite.

Then as $N_t = \sum_{l=1}^{L_t} A_{tl}$, the distribution of N_t, defined as $\{\lambda_t(n)\}_{n \geq 0}$, also follows Poisson distribution with the expectation $\mathfrak{N}_t = L_t \Lambda$.

2.3.3   Packet Success Probability

The process of randomly and independently choosing a time-slot in a frame to transmit packets can be cast into a class of problems that are known as occupancy problems, or bins and balls problem [15]. Specifically, consider the setting where a number of balls are randomly and independently placed into a number of bins, the classical occupance problem studies the maximum load of an individual bin.

In our context, time-slots and packets to be transmitted in a frame can be cast into bins and balls, respectively. Denote by Y_t the random variable for the number of packets to be transmitted in frame t. Given $Y_t = \hat{h}$ in frame t and the frame length L_t, the number x_0 of packets sent in one time-slot, referred as to occupancy number, is binomially distributed with parameters $\hat{h}$ and $\frac{1}{L_t}$:

$$B_{\hat{h}, \frac{1}{L_t}}(x_0) = \binom{\hat{h}}{x_0}(\frac{1}{L_t})^{x_0}(1 - \frac{1}{L_t})^{\hat{h} - x_0}. \tag{2.3}$$

Applying the distribution of Eq. (2.3) to all L_t slots in the frame, we can get the expected value $b(x_0)$ of the number of time-slots with occupance number x_0 in a frame as follows:

$$b(x_0) = L_t B_{\hat{h}, \frac{1}{L_t}}(x_0) = L_t \binom{\hat{h}}{x_0}(\frac{1}{L_t})^{x_0}(1 - \frac{1}{L_t})^{\hat{h} - x_0}. \tag{2.4}$$

We further derive the probability that a packet is transmitted successfully under both SPR and MPR.

2.3.3.1   Packet Success Probability of FSA-SPR

In FSA-SPR, the number of successfully received packets equals that of time-slots with occupance number $x_0 = 1$. Following the result of [29], we can obtain the probability that under SPR there exist exactly k successful packets among $\hat{h}$ transmitted packets in the frame, denoted by $\xi_{\hat{h}k}^{SPR}$, as follows:

$$\xi_{\hat{h}k}^{SPR} = \begin{cases} \frac{\binom{L_t}{k}\binom{\hat{h}}{k}k!G(L_t - k, \hat{h} - k)}{L_t^{\hat{h}}}, & 0 < k < \min(\hat{h}, L_t) \\ \frac{\binom{L_t}{\hat{h}}\hat{h}!}{L_t^{\hat{h}}}, & k = \hat{h} \leq L_t \\ 0, & k > \min(\hat{h}, L_t) \\ 0, & k = L_t < \hat{h} \end{cases} \tag{2.5}$$

where

$$G(V, w) = V^{\hat{w}} + \sum_{t=1}^{\hat{w}} (-1)^t \prod_{j=0}^{t-1} [(\hat{w} - j)(V - j)](V - t)^{\hat{w}-t} \frac{1}{t!}$$

with $V \triangleq L_t - k$ and $\hat{w} \triangleq \hat{h} - k$.

Consequently, the expected number of successfully received packets in one frame in FSA-SPR, denoted as $r_{\hat{h}}^{SPR}$, is

$$r_{\hat{h}}^{SPR} = \sum_{k=1}^{\min(\hat{h}, L_t)} k \xi_{\hat{h}k}^{SPR} = b(1). \tag{2.6}$$

2.3.3.2   Packet Success Probability of FSA-MPR

Let occupancy numbers x_l and k_l be the number of transmitted packets and successful packets in the lth time-slot, respectively, where $l = 1, 2, \cdots, L_t$. The probability that k packets are received successfully among $\hat{h}$ transmitted packets in the frame, denoted by $\xi_{\hat{h}k}^{MPR}$, can be expressed as

$$\xi_{\hat{h}k}^{MPR} = \sum_{\sum_l x_l = \hat{h}} \sum_{\sum_l k_l = k} \prod_l \hat{\xi}_{x_l k_l} \tag{2.7}$$

We can further derive the expected number of successfully received packets in one frame as

$$r_{\hat{h}}^{MPR} = \sum_{k=1}^{\hat{h}} k \xi_{\hat{h}k}^{MPR} = L_t \sum_{x_0=1}^{\hat{h}} \sum_{k_0=1}^{x_0} B_{\hat{h}, \frac{1}{L_t}}(x_0) k_0 \hat{\xi}_{x_0 k_0}. \tag{2.8}$$

In the subsequent analysis, to make the presentation concise without introducing ambiguity, we use $\xi_{\hat{h}k}$ to denote $\xi_{\hat{h}k}^{SPR}$ in FSA-SPR and $\xi_{\hat{h}k}^{MPR}$ in FSA-MPR. The notations used in the chapter are summarized in Table 2.1.

2.4   Main Results

To streamline the presentation, we summarize the main results in this section and give the detailed proof and analysis in the subsequent sections that follow.

Aiming at studying the stability of FSA, we decompose our global objective into the following three questions, all of which are of fundamental importance both on

Table 2.1 Main notations

Symbols	Descriptions
p	Persistence probability
$\overline{M}$	Maximum MPR capacity
Λ	Expected arrival rate per slot
$\mathfrak{N}_t$	Expected arrival rate in frame t
$\lambda_t(n)$	Prob. of n new arrivals in frame t
L_t	The length of frame t
X_t	Random variable: No. of backlogs in frame t
i	The value of backlogs in frame t, i.e., $X_t = i$
Y_t	Random variable: No. of transmitted packet in frame t
$\hat{h}$	The value of packets sent in frame t, i.e., $Y_t = \hat{h}$
Z_t	Random variable: No. of retransmitted packet in frame t
h	The value of retransmitted packets in frame t, i.e., $Z_t = h$
α	The ratio of $\hat{h}$ to L_t
$\hat{\xi}_{x_0 k_0}$	Prob. of having k_0 out of x_0 successful packets in a slot
$\xi_{\hat{h}k}$	Prob. of having k out of $\hat{h}$ successful packets in frame t
P_{is}	One-step transition probability
D_i	Drift in frame t

the theoretical characterisation of FSA performance and its effective operation in practical systems:

- **Q1**: Under what condition(s) is FSA stable?
- **Q2**: When is the stability region maximised?
- **Q3**: How does FSA behave in the instability region?

Before answering the questions, we first introduce the formal definition of stability employed by Ghez et al. in [19].

Define by random variable X_t the number of backlogged packets in the system at the start of frame t. The discrete-time process $(X_t)_{t \geq 0}$ can be seen as a homogeneous Markov chain.

Definition 2.1 An FSA system is stable if $(X_t)_{t \geq 0}$ is ergodic and unstable otherwise.

By Definition 2.1, we can transform the study of stability of FSA into investigating the ergodicity of the backlog Markov chain. The rationality of this transformation is twofold. One interpretation is the property of ergodicity that there exists a unique stationary distribution of a Markov chain if it is ergodic. The other can be interpreted from the nature of ergodicity that each state of the Markov chain can recur in finite time with probability 1.

From an engineering perspective, if FSA is stable, then the number of backlogs in the system will reduce overall; otherwise, it will increase as the system operates.

We then establish the following results characterizing the stability region and demonstrating the behavior of the Markov chain in nonergodicity regions under both SPR and MPR.

2.4.1 Results for FSA-SPR

Denote by i and $\hat{h}$ the value of the number of backlogs and sent packets in frame t and $\alpha \triangleq \frac{\hat{h}}{L_t}$. Recall the definitions of X_t and Y_t, we can suppose that $X_t = i$ and $Y_t = \hat{h}$.

Theorem 2.1 *Under FSA-SPR, consider an irreducible and aperiodic backlog Markov chain $(X_t)_{t \geq 0}$ with nonnegative integers. When $i \to \infty$, we have* [1]

1. *The system is always stable if $\Lambda < \alpha e^{-\alpha}$ and $L_t = \Theta(\hat{h})$. Specially, $\alpha = 1$ maximizes the stability region[2] and also the stable throughput.*
2. *The system is unstable under each of the following three conditions: (1) $L_t = o(\hat{h})$; (2) $L_t = O(\hat{h})$; (3) $L_t = \Theta(\hat{h})$ and $\Lambda > \alpha e^{-\alpha}$.*

Remark Theorem 2.1 answers the first two questions and can be interpreted as follows:

- When $L_t = o(\hat{h})$, i.e., the number of sent packets $\hat{h}$ is far larger than the frame length L_t, a packet experiences collision with high probability (w.h.p.), thus increasing the backlog size and destabilising the system;
- When $L_t = O(\hat{h})$, i.e., the number of sent packets $\hat{h}$ is far smaller than the frame length, a packet is transmitted successfully w.h.p.. However, the expected number of successful packets is still significantly less than that of new arrivals in the frame. The system is thus unstable.
- When $L_t = \Theta(\hat{h})$, i.e., $\hat{h}$ has the same order of magnitude with the frame length, the system is stable when the backlog can be reduced gradually, i.e., when the expected arrival rate is less than the successful rate.

It is well known that an irreducible aperiodic Markov chain falls into one of three mutually exclusive classes: positive recurrent, null recurrent and transient. So, our

---

[1]For two variables X, Y, we use the following asymptotic notations:

- $X = o(Y)^*$ if $0 \leq \frac{X}{Y} \leq \theta_0$, as $Y \to \infty$, where constant $\theta_0 \geq 0$;
- $X = o(Y)$ if $\frac{X}{Y} = 0$, as $Y \to \infty$;
- $X = O(Y)$ if $\frac{X}{Y} = \infty$, as $Y \to \infty$;
- $X = \Theta(Y)$ if $\theta_1 \leq \frac{X}{Y} \leq \theta_2$, as $Y \to \infty$, where constants $\theta_2 \geq \theta_1 > 0$.

[2]The ergodicity region of a Markov chain in this chapter is referred to as stability region.

next step after deriving the stability conditions is to show whether the backlog Markov chain in the instability region is transient or recurrent, which answers the third question.

Theorem 2.2 *With the same notations as in Theorem 2.1, $(X_t)_{t \geq 0}$ is always transient in the instability region, i.e., under each of the following three conditions: (1) $L_t = o(\hat{h})$; (2) $L_t = \Theta(\hat{h})$ and $\Lambda > \alpha e^{-\alpha}$; (3) $L_t = O(\hat{h})$.*

Remark If a state of a Markov chain is transient, then the probability of returning to itself for the first time in a finite time is less than 1. Hence, Theorem 2.2 implies that once out of the stability region, the system is not guaranteed to return to stable state in finite time, that is, the number of backlogs will increase persistently.

2.4.2 Results for FSA-MPR

Theorem 2.3 *Under FSA-MPR, using the same notations as in Theorem 2.1, we have*

1. *The system is always stable if $L_t = \Theta(\hat{h})$ and $\Lambda < \sum_{x_0=1}^{\overline{M}} e^{-\alpha} \frac{\alpha^{x_0}}{x_0!} \sum_{k_0=1}^{x_0} k_0 \hat{\xi}_{x_0 k_0}$. Specially, let α^* denote the value of α that maximises the upper bound of stability region, it holds that $\alpha^* = \Theta(\overline{M})$.*
2. *The system is unstable under each of the following conditions: (1) $L_t = o(\hat{h}^{1-\epsilon_1})$ where $0 < \epsilon_1 \leq 1$; (2) $L_t = O(\hat{h})$; (3) $\Lambda > \alpha$ and $L_t = \Theta(\hat{h})$.*

Remark Comparing the results of Theorem 2.3 to Theorem 2.1, we can quantify the performance gap between FSA-SPR and FSA-MPR in terms of stability. For example, when $\alpha = 1$, the stability region is maximised in FSA-SPR with $\Lambda < e^{-1}$, while the upper bound of the stability region in FSA-MPR is $e^{-1} \sum_{x_0=1}^{\overline{M}} \frac{1}{(x_0-1)!}$. Note that for $\overline{M} > 2$, it holds that

$$1 + 1 + \frac{1}{2} < \sum_{x_0=1}^{\overline{M}} \frac{1}{(x_0-1)!} < 1 + 1 + \sum_{x_0=1}^{\overline{M}} \frac{1}{x_0(x_0+1)} < 2 + \left(\sum_{x_0=1}^{\overline{M}} \frac{1}{x_0} - \frac{1}{x_0+1} \right)$$

$$= 3 - \frac{1}{\overline{M}+1}.$$

The upper bound of the stability region of FSA-MPR when $\alpha = 1$ is thus between 2.5 and 3 times the maximum stability region of FSA-SPR. And hence the maximum upper bound of the stability region of FSA-MPR achieved when $\alpha^* = \Theta(\overline{M})$ is far larger than that of FSA-SPR.

Theorem 2.4 *With the same notations as in Theorem 2.3, $(X_t)_{t \geq 0}$ is transient under each of the following three conditions: (1) $L_t = o(\hat{h}^{1-\epsilon_1})$; (2) $L_t = O(\hat{h})$; (3) $\Lambda > \alpha$ and $L_t = \Theta(\hat{h})$.*

Remark Theorem 2.4 demonstrates that despite the gain on the stability region size of FSA-MPR over FSA-SPR, their behaviors in the unstable region are essentially the same.

2.5 Stability Analysis of FSA-SPR

In this section, we will analyse the stability of FSA-SPR and prove Theorems 2.1 and 2.2.

2.5.1 Characterising Backlog Markov Chain

As mentioned in Sect. 2.4, we characterize the number of the backlogged packets in the system at the beginning of frame t as a homogeneous Markov chain $(X_t)_{t \geq 0}$. We assume that $X_t = i$ and $Y_t = \hat{h}$. Denote by Z_t the random variable for the number of retransmitted packets in frame t. Since the transmitted packets in frame t consists of the new arrivals during frame $t - 1$ and the retransmitted packets in frame t, we have

$$Y_t = Z_t + N_{t-1}. \tag{2.9}$$

Suppose w new packets arrive in frame $t - 1$ and h out of $i - w$ backlogs are retransmitted in frame t of which the probability is as follows:

$$B_{i-w}(h) \triangleq \binom{i - w}{h} p^h (1 - p)^{i-w-h}.$$

As a consequence, the number of packets transmitted in frame t is $\hat{h} = w + h$.

We now calculate the one-step transition probability as a function of $\xi_{\hat{h}k}$, retransmission probability p and $\{\lambda_t(n)\}_{n \geq 0}$. Denote by $P_{is} = P\{X_{t+1} = s | X_t = i\}$ the one-step transition probability, we can derive the following results:

1. For $i = 0$:

$$P_{00} = \lambda_t(0),$$

$$P_{0s} = \lambda_t(s), \quad s \geq 1,$$

2. for $i \geq 1$:

$$
\begin{cases}
P_{i,i-s} = \sum_{w=0}^{i} \lambda_{t-1}(w) \sum_{h=\{s-w\}^+}^{i-w} B_{i-w}(h) \cdot \\
\qquad \sum_{n=0}^{\min(L,\hat{h})-s} \lambda_t(n) \xi_{\hat{h},n+s}, \qquad 1 \leq s \leq i, \\
P_{i,i} = \lambda_t(0)\Big(\lambda_{t-1}(0) B_i(0) + \sum_{w=0}^{i} \lambda_{t-1}(w) \cdot \\
\qquad \sum_{h=0}^{i-w} B_{i-w}(h) \xi_{\hat{h},0}\Big) + \sum_{w=0}^{i} \lambda_{t-1}(w) \cdot \\
\qquad \sum_{h=0}^{i-w} B_{i-w}(h) \sum_{n=1}^{\min(\hat{h},L)} \lambda_t(n) \xi_{\hat{h}n}, \\
P_{i,i+s} = \sum_{w=0}^{i} \lambda_{t-1}(w) \sum_{h=0}^{i-w} B_{i-w}(h) \cdot \\
\qquad \sum_{n=0}^{\min(\hat{h},L)} \lambda_t(n+s) \xi_{\hat{h}n}, \qquad s \geq 1,
\end{cases}
\tag{2.10}
$$

where $\{s - w\}^+ = \max\{s - w, 0\}$.

The rationale for the calculation of the transition probability is explained as follows:

- When $i = 0$, i.e., there are no backlogs in the frame, the backlog size remains zero if no new packets arrive and increases by s if s new packets arrive in the frame.
- When $i > 0$, we have three possibilities, corresponding to the cases where the backlog size decreases, remains unchanged and increases, respectively:
  - The state $1 \leq s \leq \min(\hat{h}, L_t)$ corresponds to the case where the backlog size decreases by s when $n \leq \min(\hat{h}, L_t) - s$ new packets arrive but $n + s$ backlogged packets are received successfully.
  - The backlog size remains unchanged if either of two following events happens: (a) no new packets are generated and either no backlogged packets are transmitted or all the transmitted backlogged packets fail; (b) $n \leq \min(\hat{h}, L_t)$ new packets arrive but n backlogged packets are successfully received.
  - The backlog size increases when the number of successful packets is less than that of new arrivals.

In order to establish the ergodicity of the backlog Markov chain $(X_t)_{t\geq 0}$, it is necessary to ensure $(X_t)_{t\geq 0}$ is irreducible and aperiodic. To this end, we conclude this subsection by providing the sufficient conditions on $\{\lambda_t(n)\}$ for the irreducibility and the aperiodicity of $(X_t)_{t\geq 0}$ as

$$
0 < \lambda_t(n) < 1, \ \forall n \geq 0.
\tag{2.11}
$$

We would like to point out that most of traffic models can satisfy (2.11). Throughout the chapter, it is assumed that (2.11) holds and hence $(X_t)_{t\geq 0}$ is irreducible and aperiodic.

2.5.2 Stability Analysis

Recalling Definition 2.1, to study the stability of FSA, we need to analyse the ergodicity of the backlog Markov chain $(X_t)_{t\geq 0}$. We first define the drift and then introduce two auxiliary lemmas which will be useful in the ergodicity demonstration.

Definition 2.2 The drift D_i of the backlog Markov chain $(X_t)_{t\geq 0}$ at state $X_t = i$ where $i \geq 0$ is defined as

$$D_i = E[X_{t+1} - X_t | X_t = i]. \tag{2.12}$$

Lemma 2.1 ([30]) *Given an irreducible and aperiodic Markov chain $(X_t)_{i\geq 0}$ having nonnegative integers as state space with the transition probability matrix $\mathbf{P} = \{P_{is}\}$, $(X_t)_{t\geq 0}$ is ergodic if for some integer $Q \geq 0$ and constant $\epsilon_0 > 0$, it holds that*

1. $|D_i| < \infty$, for $i \leq Q$,
2. $D_i < -\epsilon_0$, for $i > Q$.

Lemma 2.2 ([31]) *Under the assumptions of Lemma 2.1, $(X_t)_{t\geq 0}$ is not ergodic, if there exist some integer $Q \geq 0$ and some constants $B \geq 0$, $c \in [0, 1]$ such that*

1. $D_i > 0$ for all $i \geq Q$,
2. $\phi^i - \sum_s P_{is}\phi^i \geq -B(1 - \phi)$ for all $i \geq Q$, $\phi \in [c, 1]$.

Armed with Lemma 2.1 and Lemma 2.2, we start to prove Theorem 2.1.

Proof of Theorem 2.1 In the proof, we first explicitly formulate the drift defined by (2.12) and then study the ergodicity of Markov chain based on drift analysis.

Denote by random variable C_t the number of successful transmissions in frame t, we have

$$X_{t+1} - X_t = N_t - C_t.$$

Recall (2.12), it then follows that

$$D_i = E[N_t - C_t | X_t = i] = \mathfrak{N}_t - E[C_t | X_t]. \tag{2.13}$$

Since all new arrivals and unsuccessful packets in frame $t - 1$ are transmitted in frame t with probability one and p, respectively, we have

$$P\{C_t = k | X_t = i, N_{t-1} = w, Z_t = h\} = \xi_{\hat{h}k}^{SPR},$$

for $0 \leq k \leq \min(\hat{h}, L)$. Recall (2.6), we have

$$E[C_t|X_t = i] = \sum_{h=0}^{i-w} B_{i-w}(h) E[C_t = k|X_t = i, N_{t-1} = w]$$

$$= \sum_{w=0}^{i} \lambda_{t-1}(w) \sum_{h=0}^{i-w} B_{i-w}(h) r_{\hat{h}}^{SPR}. \tag{2.14}$$

Following (2.13) and (2.14), we obtain the value of the drift as follows:

$$D_i = \mathfrak{N}_t - \sum_{w=0}^{i} \lambda_{t-1}(w) \sum_{h=0}^{i-w} B_{i-w}(h) r_{\hat{h}}^{SPR}. \tag{2.15}$$

After formulating the drift, we then proceed by two steps.

Step 1: $L_t = \Theta(\hat{h})$ and $\Lambda < \alpha e^{-\alpha}$

In this step, we intend to corroborate that the conditions in Lemma 2.1 can be satisfied if $L_t = \Theta(\hat{h})$ and $\Lambda < \alpha e^{-\alpha}$. We first show that $|D_i|$ is finite. This is true for $i \leq Q$ since

$$|D_i| < \max\{\mathfrak{N}_t, \sum_{w=0}^{i} \lambda_{t-1}(w) \sum_{h=0}^{i-w} B_{i-w}(h) r_{\hat{h}}^{SPR}\}$$

$$< \max\{\mathfrak{N}_t, \min\{L_t, \sum_{w=0}^{i} \lambda_{t-1}(w) \sum_{h=0}^{i-w} (w + h) B_{i-w}(h)\}\}$$

$$< \max\{\mathfrak{N}_t, \min\{L_t, \sum_{w=0}^{i} w\lambda_{t-1}(w) + \sum_{w=0}^{i} (i - w) p\lambda_{t-1}(w)\}\}$$

$$< \max\{\mathfrak{N}_t, \min\{L_t, (1 - p)\lambda + ip\}\}. \tag{2.16}$$

Next, to derive the limit of D_i, we start with the following lemma which is proved in Sect. 2.10.1.

Lemma 2.3 *If $r_{\hat{h}}^{SPR}$ has a limit $\hat{r}$, then it holds that*

$$\lim_{i \to \infty} \sum_{w=0}^{i} \lambda_{t-1}(w) \sum_{h=0}^{i-w} B_{i-w}(h) r_{\hat{h}}^{SPR} = \hat{r}.$$

Following Lemma 2.3, we have

$$\lim_{i \to \infty} D_i = \mathfrak{N}_t - \lim_{\hat{h} \to \infty} r_{\hat{h}}^{SPR}$$

$$= \lim_{\hat{h} \to \infty} L_t \left\{ \Lambda - \binom{\hat{h}}{1} \frac{1}{L_t} \cdot \left(1 - \frac{1}{L_t} \right)^{\hat{h}-1} \right\}$$

$$= L_t (\Lambda - \alpha e^{-\alpha}), \tag{2.17}$$

where $\alpha \triangleq \frac{\hat{h}}{L_t}$. It thus holds that $\lim_{i \to \infty} D_i < -\epsilon_0$ with $\epsilon_0 = \frac{\alpha e^{-\alpha} - \Lambda}{2}$ since both α and Λ are constants when $L_t = \Theta(\hat{h})$ and $\Lambda < \alpha e^{-\alpha}$.

It then follows from Lemma 2.1 that $(X_t)_{t \geq 0}$ is ergodic. Specially, when $\alpha = 1$, the system stability region is maximized, i.e., $\Lambda < e^{-1}$.

Step 2: $L_t = o(\hat{h})$ **or** $L_t = O(\hat{h})$ **or** $L_t = \Theta(\hat{h})$ **and** $\Lambda > \alpha e^{-\alpha}$

In this step, we prove the instability of $(X_t)_{t \geq 0}$ by applying Lemma 2.2. Taking into consideration the impact of different relation between L_t and $\hat{h}$ on the limit of D_i. With (2.17), the following results hold for $\hat{h} \to \infty$:

- $\Lambda - \lim_{\alpha \to \infty} \alpha e^{-\alpha} = \Lambda > 0$, when $L_t = o(\hat{h})$,
- $\Lambda - \lim_{\alpha \to 0} \alpha e^{-\alpha} = \Lambda > 0$, when $L_t = O(\hat{h})$,
- $\Lambda - \alpha e^{-\alpha} > 0$, when $L_t = \Theta(\hat{h})$ and $\Lambda > \alpha e^{-\alpha}$.

Consequently, we have $\lim_{i \to \infty} D_i > 0$, which proves the first condition in Lemma 2.2.

Next, we will validate the second condition of Lemma 2.2 in two cases according to the probable relationship between $\hat{h}$ and i, i.e., $\hat{h} = o(i)$ and $\hat{h} = \Theta(i)$.

Note that the second condition apparently holds for $\phi = 0$ and $\phi = 1$, we thus focus on the remaining value of ϕ, i.e., $\phi \in (c, 1)$. Moreover, given $\hat{h}$, $P_{i,i-s}$ in (2.10) can also be expressed as

$$P_{i,i-s} = \sum_{w=0}^{\hat{h}} \lambda_{t-1}(w) B_{i-w}(\hat{h} - w) \sum_{n=0}^{\hat{h}-s} \lambda_t(n) \xi_{\hat{h}, n+s}. \tag{2.18}$$

Now, we start the proof with the above arms.

Case 1: $\hat{h} = o(i)$

Given $\hat{h} = o(i)$, we can derive the result as follows:

$$\sum_{s=0}^{\infty} \phi^s P_{is} = \sum_{s=0}^{i-\hat{h}-1} \phi^s P_{is} + \sum_{s=i-\hat{h}}^{i} \phi^s P_{is} + \sum_{s=i+1}^{\infty} \phi^s P_{is}$$

$$\leq \phi^{i+1} + \sum_{s=i-\hat{h}}^{i} \phi^s \sum_{w=0}^{\hat{h}} \lambda_{t-1}(w) B_{i-w}(\hat{h} - w) \cdot \sum_{n=0}^{\hat{h}+s-i} \lambda_t(n) \xi_{\hat{h}, n+i-s}$$

$$\leq \phi^{i+1} + \sum_{s=i-\hat{h}}^{i} \phi^s \sum_{w=0}^{\hat{h}} B_{i-w}(\hat{h} - w)$$

$$\leq \phi^{i+1} + \hat{h} e^{-\frac{ip}{2}(1-\frac{\hat{h}}{ip})^2} \phi^{i-\hat{h}} \leq \phi^i, \text{ as } i \to \infty, \tag{2.19}$$

where we use the Chernoff's inequality to bound the cumulative probability of $B_{i-w}(\hat{h} - w)$. Therefore, the second condition of Lemma 2.2 holds when $\hat{h} = o(i)$.

Case 2: $\hat{h} = \Theta(i)$

In this case, we need to distinguish the three instability regions. Without loss of generality, we assume that $\hat{h} = \beta i$ where constant $\beta \in (0, 1]$.

(1) $L_t = o(\hat{h})$.

When $L_t = o(\hat{h})$, it also holds that $L_t = o(i)$ and that at most $L_t - 1$ packets are successfully received, we thus have

$$\sum_{s=0}^{\infty} \phi^s P_{is} = \sum_{s=0}^{i-L_t-1} \phi^s P_{is} + \sum_{s=i-L_t}^{i} \phi^s P_{is} + \sum_{s=i+1}^{\infty} \phi^s P_{is}$$

$$\leq \phi^{i+1} + L_t \phi^{i-L_t} \leq \phi^i + B(1 - \phi), \text{ as } i \to \infty, \tag{2.20}$$

for any positive constant B.

(2) $L_t = \Theta(\hat{h})$.

The key steps we need in this case are to obtain upper bounds of $\xi_{\hat{h}k}^{SPR}$ and the arrival rate in a new way. To this end, we first recomputed $\xi_{\hat{h}k}^{SPR}$ when $L_t = \Theta(\hat{h})$ as follows:

$$\begin{cases} \xi_{\hat{h}k}^{SPR} = \dfrac{\binom{L_t}{k}k!(L_t-k)^{\hat{h}-k}}{L_t^{\hat{h}}} - \dfrac{\binom{L_t}{k+1}(k+1)!(L_t-k-1)^{\hat{h}-k-1}}{L_t^{\hat{h}}} \\[2ex] \quad\; \leq \dfrac{\binom{L_t}{k}k!}{L_t^{k}}\left((1 - \tfrac{k}{L_t})^{\hat{h}-k} - (1 - \tfrac{k+1}{L_t})^{\hat{h}-k}\right) \\[2ex] \quad\; \leq \dfrac{\binom{L_t}{k}k!}{L_t^{k}} \\[2ex] \quad\; \leq (1 - \tfrac{k/2}{L_t})^{k/2}, \qquad k < \min(\hat{h}, L_t), \\[2ex] \xi_{\hat{h}\hat{h}}^{SPR} = \dfrac{\binom{L_t}{\hat{h}}\hat{h}!}{L_t^{\hat{h}}} \\[2ex] \quad\; \leq (1 - \tfrac{\hat{h}/2}{L_t})^{\hat{h}/2} \\[2ex] \quad\; \leq (1 - \alpha/2)^{\hat{h}/2}, \qquad k = \hat{h} \leq L_t. \end{cases} \tag{2.21}$$

The rationale behind the above inequalities is as follows: Given $\hat{h}$ transmitted packets, the probability of exactly k successful packets equals the absolute value of the difference between the probability of at least k successful packets and that of at least $k + 1$ successful packets.

Next, we introduce an auxiliary lemma to bound the probability distribution of the arrival rate. When the number of new arrivals per slot A_{tl} is Poisson distributed with the mean Λ, the number of new arrivals per frame N_t (A_{tl} and N_t is formally defined in Sect. 2.3.) is also a Poisson random variable with the mean $\mathfrak{N}_t = L_t \Lambda > \hat{h} e^{-\alpha}$.

Lemma 2.4 ([32]) *Given a Poisson distributed variable X with the mean μ, it holds that*

$$Pr[X \le x] \le \frac{e^{-\mu}(e\mu)^x}{x^x}, \quad \forall\, x < \mu, \tag{2.22}$$

$$Pr[X \ge x] \le \frac{e^{-\mu}(e\mu)^x}{x^x}, \quad \forall\, x > \mu. \tag{2.23}$$

In the case that $L_t = \Theta(i)$, it holds that $\mathfrak{N}_t = L_t \Lambda > L_t^{\frac{2}{3}}$, for the constant Λ and a large i. Consequently, applying (2.22) in Lemma 2.4, we have

$$P\{N_t \le L_t^{2/3}\} \le \frac{e^{-\lambda}(e\lambda)^{L_t^{2/3}}}{(L_t^{2/3})^{L_t^{2/3}}} \le e^{-L_t^{2/3}(\frac{L_t\Lambda}{L_t^{2/3}}-1)} \left(\frac{L_t\Lambda}{L_t^{2/3}}\right)^{L_t^{2/3}}$$

$$\le \left(e^{\frac{L_t\Lambda}{L_t^{2/3}}-1}\frac{L^{2/3}}{L_t\Lambda}\right)^{-L_t^{2/3}} \le \frac{1}{a_1^{L_t^{2/3}}}, \tag{2.24}$$

where $a_1 \triangleq \frac{e^{\Lambda L_t^{1/3}-1}}{\Lambda L_t^{1/3}} \gg 1$, following the fact that $e^x > 1 + x$, for $\forall\, x > 0$.

Armed with (2.21) and (2.24) and noticing the fact that at most $\hat{h}$ packets are successfully received, we start developing the proof and obtain the results as follows:

$$\sum_{s=0}^{\infty} \phi^s P_{is} = \sum_{s=0}^{i-\hat{h}-1} \phi^s P_{is} + \sum_{s=i-\hat{h}}^{i} \phi^s P_{is} + \sum_{s=i+1}^{\infty} \phi^s P_{is}$$

$$\le \phi^{i+1} + \sum_{s=i-\hat{h}}^{i} \sum_{w=0}^{\hat{h}} \lambda_{t-1}(w) B_{i-w}(\hat{h} - w)$$

$$\cdot \sum_{n=0}^{\hat{h}+s-i} \lambda_t(n)(1 - \frac{n+i-s}{2L_t})^{(n+i-s)/2}$$

$$\le \phi^{i+1} + \sum_{s=i-\hat{h}}^{i} \phi^s \sum_{n=0}^{\hat{h}} \lambda_t(n)(1 - \frac{n}{2L_t})^{\frac{n}{2}}$$

$$\le \phi^{i+1} + \sum_{s=i-\hat{h}}^{i} \phi^s \left(\sum_{n=0}^{L_t^{2/3}} \lambda_t(n) + \sum_{n=L_t^{2/3}}^{\hat{h}} \lambda_t(n)(1 - \frac{n}{2L_t})^{\frac{n}{2}}\right)$$

$$\leq \phi^{i+1} + \sum_{s=i-\hat{h}}^{i} \phi^s \left(\frac{1}{a_1^{\frac{L_t^{2/3}}{2}}} + (1 - \frac{1}{2L_t^{1/3}})^{\frac{L_t^{2/3}}{2}} \right)$$

$$\leq (\hat{h} + 1)\left(\frac{1}{a_1^{\frac{L_t^{2/3}}{2}}} + e^{-\frac{L_t^{1/3}}{4}} \right) \phi^{i-\hat{h}} + \phi^{i+1} \leq \phi^i + B(1 - \phi), \quad \text{as } i \to \infty,$$

$$(2.25)$$

for any positive constant B, where the last inequality holds for $(\hat{h} + 1)\left(\frac{1}{a_1^{\frac{L_t^{2/3}}{2}}} + e^{-\frac{L_t^{1/3}}{4}} \right) \sim \Theta(ie^{-i^{1/3}}) \to 0$ as $i \to \infty$, while $B(1 - \phi)$ is positive constant.

Consequently, the second condition in Lemma 2.2 holds for Case 2. Next, we proceed with the proof for the third case.

(3) $L_t = O(\hat{h})$.

When $L_t = O(\hat{h})$, it also holds that $L_t = O(i)$ such that the expected number of new arrivals per frame $\mathfrak{N}_t = L_t \Lambda \gg i$. Since N_t is Poisson distributed as mentioned in Case 2 above, recall (2.24), it also holds that

$$P\{N_t \leq i\} \leq \frac{1}{a_2^i}, \qquad (2.26)$$

where $a_2 \triangleq \frac{i}{L_t \Lambda} \cdot e^{\frac{L_t \Lambda}{i} - 1} \geq \frac{L_t \Lambda}{i}$, following the fact that $e^x > 1 + x + \frac{x^2}{2} + \frac{x^3}{6}$, for $\forall x > 0$.

Using (2.26) then yields

$$\sum_{s=0}^{\infty} \phi^s P_{is} = \sum_{s=0}^{i} \phi^s P_{is} + \sum_{s=i+1}^{\infty} \phi^s P_{is} \leq \sum_{s=0}^{i} \phi^s \sum_{n=0}^{s} \lambda_t(n) + \phi^{i+1}$$

$$\leq \sum_{s=0}^{i} \sum_{n=0}^{s} \lambda_t(n) + \phi^{i+1} \leq \frac{i+1}{(\phi \frac{L_t \Lambda}{i})^i} \phi^i + \phi^{i+1} \leq \phi^i, \quad \text{as } i \to \infty,$$

$$(2.27)$$

since ϕ is constant while $\frac{L_t \Lambda}{i} \to \infty$ as $i \to \infty$.

Combining the analysis above, it follows Lemma 2.2 that the backlog Markov chain $(X_t)_{t \geq 0}$ is unstable when $L_t = o(\hat{h})$ or $L_t = O(\hat{h})$ or $L_t = \Theta(\hat{h})$ and $\Lambda > \alpha e^{-\alpha}$. And the proof of Algorithm 2.1 is thus completed. □

2.5.3 System Behavior in Instability Region

It follows from Theorem 2.1 that the system is unstable in the following three conditions: $L_t = o(\hat{h})$; $L_t = O(\hat{h})$; and $L_t = \Theta(\hat{h})$ but $\Lambda > \alpha e^{-\alpha}$. Lemma 2.2, however, is not sufficient to ensure the transience of a Markov chain, we thus in this section further investigate the system behavior in the instability region, i.e., when $(X_t)_{t\geq 0}$ is nonergodic. The key results are given in Theorem 2.2.

Before proving Theorem 2.2, we first introduce the following lemma [33] on the conditions for the transience of a Markov chain.

Lemma 2.5 ([33]) *Let $(X_t)_{t\geq 0}$ be an irreducible and aperiodic Markov chain with the nonnegative integers as its state space and one-step transition probability matrix $\mathbf{P} = \{P_{is}\}$. $(X_t)_{t\geq 0}$ is transient if and only if there exists a sequence $\{y_i\}_{i\geq 0}$ such that*

1. y_i $(i \geq 0)$ is bounded,
2. for some $i \geq N$, $y_i < y_0$, y_1, $\cdots$, y_{N-1},
3. for some integer $N > 0$, $\sum_{s=0}^{\infty} y_s P_{is} \leq y_i$, $\forall\, i \geq N$.

Armed with Lemma 2.5, we now prove Theorem 2.2.

Proof of Theorem 2.2 The key to prove Theorem 2.2 is to show the existence of a sequence satisfying the properties listed in Lemma 2.5, so we first construct the following sequence (2.28) and then prove that it satisfies the required conditions.

$$y_i = \frac{1}{(i+1)^{\theta}}, \quad \theta \in (0, 1). \tag{2.28}$$

It can be easily checked that $\{y_i\}$ satisfies the first two properties in Lemma 2.5.

Noticing that the sequence $\{\phi^i\}$ in Lemma 2.2 satisfies the first two properties in Lemma 2.5 for $0 < \phi < 1$, and recall (2.19) and (2.27), we can conclude that $(X_t)_{t\geq 0}$ is transient if $\hat{h} = o(i)$ or $L_t = O(i)$. Therefore, we next proceed with $\hat{h} = \Theta(i)$ by distinguish two cases.

Case 1: $L_t = o(\hat{h})$

When $\hat{h} = \Theta(i)$, it also holds that $L_t = o(i)$. To streamline the complicated analysis in this case, we partition the region $L_t = o(i)$ into two parts, i.e., (1) $L_t = o((\ln i)^4)^*$, and (2) $L_t = o(i)$ except part (1), i.e., the region $[O((\ln i)^4), o(i)]$.

- Part (1): $L_t = o((\ln i)^4)^*$.

  The result in this part is shown in the following lemma for the third property in Lemma 2.5. The proof is detailed in Sect. 2.10.2.

  Lemma 2.6 *If $L_t = o((\ln i)^4)^*$, then $(X_t)_{t\geq 0}$ is always transient.*

- Part (2): $L_t = o(i)$ except part (1).

  In this case, since $a_1 > \ln i$ and $y_i - y_{i+1} = \frac{1}{(i+1)^{\theta}}(1 - (1 - \frac{1}{i+2})^{\theta}) \geq \frac{\theta}{(i+1)^{\theta}(i+2)}$ where we use the fact that $(1 - \frac{1}{i+2})^{\theta} \leq 1 - \frac{\theta}{i+2}$ following Taylor's theorem, using (2.24) and (2.28) yields

$$\sum_{s=0}^{\infty} y_s P_{is} = \sum_{s=0}^{i-L_t} y_s P_{is} + \sum_{s=i-L_t+1}^{i} y_s P_{is} + \sum_{s=i+1}^{\infty} y_s P_{is}$$

$$\leq y_{i+1} + \sum_{s=i-L_t+1}^{i} y_s \sum_{n=0}^{s} \lambda_t(n)(1 - \frac{n}{2L_t})^{\frac{n}{2}}$$

$$\leq \sum_{s=i-L_t+1}^{i} y_s \left(\sum_{n=0}^{L_t^{\frac{2}{3}}} \lambda_n + \sum_{n=L_t^{\frac{2}{3}}+1}^{s} \lambda_t(n)(1 - \frac{n}{2L_t})^{\frac{n}{2}} \right) + y_{i+1}$$

$$\leq \sum_{s=i-L_t+1}^{i} y_s \left(\frac{1}{a_1^{L_t^{2/3}}} + (1 - \frac{1}{2L_t^{1/3}})^{\frac{L_t^{2/3}}{2}} \right) + y_{i+1}$$

$$\leq \frac{L}{(i - L_t + 2)^{\theta}} \left(\frac{1}{a_1^{L_t^{2/3}}} + e^{-\frac{L_t^{1/3}}{4}} \right) + \frac{1}{(i + 2)^{\theta}}$$

$$\leq \frac{L_t}{(i - L_t + 2)^{\theta}} \left((\ln i)^{-(\ln i)^{8/3}} + i^{-\frac{(\ln i)^{1/3}}{4}} \right) + \frac{1}{(i + 2)^{\theta}}$$

$$\leq i^{-4} + \frac{1}{(i + 2)^{\theta}} \leq \frac{1}{(i + 1)^{\theta}}, \quad \text{as } i \to \infty. \tag{2.29}$$

Case 2: $L_t = \Theta(\hat{h})$

In this case, the method to prove is similar with that used in (2.25).
Recall (2.25) , we have

$$\sum_{s=0}^{\infty} y_s P_{is} = \sum_{s=0}^{i-\hat{h}-1} y_s P_{is} + \sum_{s=i-\hat{h}}^{i} y_s P_{is} + \sum_{s=i+1}^{\infty} y_s s P_{is}$$

$$\leq y_{i+1} + \sum_{s=i-\hat{h}}^{i} y_s \left(\frac{1}{a_1^{L^{2/3}}} + (1 - \frac{1}{2L^{1/3}})^{\frac{L^{2/3}}{2}} \right)$$

$$\leq \frac{\hat{h} + 1}{(i - \hat{h} + 1)^{\theta}} \left(\frac{1}{a_1^{L^{2/3}}} + e^{-\frac{L^{1/3}}{4}} \right) + \frac{1}{(i + 2)^{\theta}} \leq \frac{1}{(i + 1)^{\theta}}, \quad \text{as } i \to \infty. \tag{2.30}$$

Consequently, it follows Lemma 2.5 that the backlog Markov chain $(X_t)_{t \geq 0}$ is transient in the instability region, which completes the proof of Theorem 2.2. $\square$

2.6 Stability Analysis of FSA-MPR

In this section, we study stability properties of FSA-MPR. Following a similar procedure as the analysis of FSA-SPR, we first establish conditions for the stability of FSA-MPR and further analyse the system behavior in the instability region.

2.6.1 Stability Analysis

We employ Lemmas 2.1 and 2.2 as mathematical base to study the stability properties of FSA-MPR, more specifically, in the proof of Theorem 2.3.

Proof of Theorem 2.3 We develop our proof in 3 steps.

Step 1: Stability Conditions

In step 1, we prove the conditions for the stability of $(X_t)_{t \geq 0}$, i.e., $\Lambda <$ $\sum_{x_0=1}^{\overline{M}} e^{-\alpha} \frac{\alpha^{x_0}}{x_0!} \sum_{k_0=1}^{x_0} k_0 \hat{\xi}_{x_0 k_0}$ and $L_t = \Theta(\hat{h})$.

Similar to (2.15), the drift at state i of $(X_t)_{t \geq 0}$ in FSA-MPR can be written as:

$$D_i = \mathfrak{N}_t - \sum_{w=0}^{i} \lambda_{t-1}(w) \sum_{h=0}^{i-w} B_{i-w}(h) r_{\hat{h}}^{MPR}. \tag{2.31}$$

According to (2.16), D_i is finite as shown in the following inequality:

$$|D_i| < \max\{\mathfrak{N}_t, (1-p)\mathfrak{N}_t + ip\},$$

which demonstrates the first conditions in Lemma 2.1 for the ergodicity of $(X_t)_{t \geq 0}$.

Recall (2.8) and Lemma 2.3, we have

$$\lim_{i \to \infty} D_i = \lim_{i \to \infty} \mathfrak{N}_t - \sum_{w=0}^{i} \lambda_{t-1}(w) \sum_{h=0}^{i-w} B_{i-w}(h) r_{\hat{h}}^{MPR}$$

$$= \lim_{\hat{h} \to \infty} L_t \left(\Lambda - \sum_{x_0=1}^{\hat{h}} B_{\hat{h}, \frac{1}{L}}(x_0) \sum_{k_0=1}^{x_0} k_0 \hat{\xi}_{x_0 k_0} \right)$$

$$= L_t \left(\Lambda - \sum_{x_0=1}^{\overline{M}} e^{-\alpha} \frac{\alpha^{x_0}}{x_0!} \sum_{k_0=1}^{x_0} k_0 \hat{\xi}_{x_0 k_0} \right). \tag{2.32}$$

Therefore, it holds that $\lim_{i \to \infty} D_i < -\epsilon_0$ if $L_t = \Theta(\hat{h})$ and $\Lambda < \sum_{x_0=1}^{\overline{M}} e^{-\alpha} \frac{\alpha^{x_0}}{x_0!} \sum_{k_0=1}^{x_0} k_0 \hat{\xi}_{x_0 k_0} \triangleq \hat{R}_1$. It then follows from Lemma 2.1 that $(X_t)_{t \geq 0}$ is ergodic with $\epsilon_0 = \frac{\hat{R}_1 - \Lambda}{2}$.

Step 2: $\alpha^* = \Theta(\overline{M})$

In Step 2, we show that $\alpha^* = \Theta(\overline{M})$. Since the proof consists mainly of algebraic operations of function optimization, we state the following lemma proving Step 2 and detail its proof in Sect. 2.10.3.

Lemma 2.7 *Let α^* denote the value of α that maximises the upper bound of the stability region, it holds that $\alpha^* = \Theta(\overline{M})$.*

Step 3: Instability Region

In Step 3, we prove the instability region of $(X_i)_{i \geq 0}$ by applying Lemma 2.2.

When $L_t = o(\hat{h})$, recall (2.32), we have

$$\sum_{x_0=1}^{\hat{h}} B_{\hat{h}, \frac{1}{L}}(x_0) \sum_{k_0=1}^{x_0} k_0 \hat{\xi}_{x_0 k_0} = \sum_{x_0=1}^{\overline{M}} B_{\hat{h}, \frac{1}{L}}(x_0) \sum_{k_0=1}^{x_0} k_0 \hat{\xi}_{x_0 k_0} \leq \sum_{x_0=1}^{\overline{M}} x_0 B_{\hat{h}, \frac{1}{L}}(x_0) \to 0,$$

$$\text{as } \hat{h} \to \infty,$$

since $\lim_{\hat{h} \to \infty} B_{\hat{h}, \frac{1}{L}}(x_0) = 0$ for a finite $\overline{M}$.

Moreover, for $L_t = O(\hat{h})$, it can be derived from (2.32) that $\sum_{x_0=1}^{\overline{M}} e^{-\alpha} \frac{\alpha^{x_0}}{x_0!} \sum_{k_0=1}^{x_0} k_0 \hat{\xi}_{x_0 k_0} \to 0$ since $\alpha \to 0$ as $\hat{h} \to \infty$.

Furthermore, according to the analysis in the first step, we know that $\lim_{i \to \infty} D_i > 0$, if the conditions in the first step are not satisfied.

Additionally, in the analysis of FSA-SPR system, we have proven that if $\hat{h} = o(i)$ or $L_t = o(\hat{h})$ or $L_t = O(\hat{h})$, the Markov chain $(X_t)_{t \geq 0}$ is always unstable, independent of $\xi_{\hat{h}k}$. Noticing that $\xi_{\hat{h}k}$ is the only difference between FSA-SPR and FSA-MPR, it thus also holds that $(X_t)_{t \geq 0}$ is unstable under FSA-MPR in the three cases.

We next study the instability of FSA-MPR when $L_t = \Theta(\hat{h})$ and $\Lambda > \alpha$. In this case, it holds that $\mathfrak{N}_t = L_t \Lambda > \hat{h}$ such that

$$P\{N_t \leq \hat{h}\} \leq \frac{1}{a_3^{\hat{h}}}, \tag{2.33}$$

where $a_3 \triangleq \frac{\alpha}{\Lambda} e^{\frac{\Lambda}{\alpha} - 1} > 1$.

Note that the one-step transition probability P_{is} in FSA-MPR can be obtained by replacing $\min(\hat{h}, L_t)$ with $\hat{h}$ in (2.10).

Hence, recall (2.25), we have

$$\sum_{s=0}^{\infty} \phi^s P_{is} = \sum_{s=0}^{i} \phi^s P_{is} + \sum_{s=i}^{\infty} \phi^s P_{is} \leq \sum_{s=0}^{i} \phi^s \sum_{n=0}^{\hat{h}} \lambda_t(n) \xi_{\hat{h}, n+i-s} + \phi^{i+1}$$

$$\leq \frac{1}{a_3^{\beta i}} + \phi^{i+1} \leq \phi^i + B(1 - \phi), \quad \text{as } i \to \infty, \tag{2.34}$$

which proves the instability of FSA-MPR following Lemma 2.2 and also completes the proof of Theorem 2.3. $\square$

2.6.2 System Behavior in Instability Region

It follows from Theorem 2.3 that the system is unstable under the following three conditions: $L_t = o(\hat{h})$; $L_t = O(\hat{h})$; $L_t = \Theta(\hat{h})$ and $\Lambda > \alpha$. In this subsection, we further investigate the system behavior in the instability region, i.e., when $(X_t)_{t \geq 0}$ is nonergodic. The key results are given in Theorem 2.4, whose proof is detailed as follows.

Proof of Theorem 2.4 In the proof of Theorem 2.2, we have proven that when $L_t = O(\hat{h})$ or $\hat{h} = o(i)$, the Markov chain $(X_t)_{t \geq 0}$ is always transient, we thus develop the proof for $\hat{h} = \Theta(i)$ by distinguishing two cases.

Case 1: $L_t = o(\hat{h}^{1-\epsilon_1})^*$ **with** $\epsilon_1 \in (0, 1]$

In this case, it holds that $L_t = o(i^{1-\epsilon_1})^*$ for $\hat{h} = \Theta(i)$. As counterparts in FSA-SPR, we also partition the region into two parts, i.e., 1) $L_t = o((\ln i)^4)^*$, and 2) $L_t = o(i)$ except part 1), i.e., the region $[O((\ln i)^4), o(i)]$.

Recall the proof of Lemma 2.6, it has been shown that $(X_t)_{t \geq 0}$ is always transient, independent of $\xi_{\hat{h}k}$, meaning $(X_t)_{t \geq 0}$ is also transient in FSA-MPR when $L_t = o((\ln i)^4)^*$.

As a consequence, it is sufficient to show the transience of $(X_t)_{t \geq 0}$ in part 2). The key step here is to obtain the upper bound of $\xi_{\hat{h}k}$. To this end, we first introduce the following auxiliary lemma.

Lemma 2.8 ([34]) *Given $\hat{h}$ packets, each packet is sent in a slot picked randomly among L_t time-slots in frame t. If $\rho_j = L_t \frac{e^{-\hat{h}/L_t}}{j!}(\frac{\hat{h}}{L_t})^j$ remains bounded for $\hat{h}, L_t \to \infty$, then the probability $P(m_j)$ of finding exactly m_j time-slots with j packets can be approximated by the following Poisson distribution with the parameter ρ_j,*

$$P(m_j) = e^{-\rho_j} \frac{\rho_j^{m_j}}{m_j!}. \tag{2.35}$$

We next show that Lemma 2.8 is applicable to FSA-MPR when $L_t = o(\hat{h}^{1-\epsilon_1})^*$ for a large enough $\hat{h}$. To that end, we verify the boundedness of ρ_j, which is derived as

$$0 \leq \rho_j \leq \frac{\hat{h}^j}{j! L_t^{j-1} e^{\hat{h}\epsilon}} \leq \frac{\hat{h}^j}{j! L_t^{j-1}} \cdot \frac{(\lceil \frac{1}{\epsilon} \rceil j)!}{(\hat{h}^\epsilon)^{\lceil \frac{1}{\epsilon} \rceil j}} \leq \frac{(\lceil \frac{1}{\epsilon} \rceil j)!}{j! L_t^{j-1}}, \tag{2.36}$$

meaning that ρ_j is bounded if j is finite.

Apparently, when $L_t = o(\hat{h}^{1-\epsilon_1})^*$, the probability of finding exactly m_j time-slots with j packets in FSA-MPR can be approximated by the Poisson distribution with the parameter ρ_j, following from Lemma 2.8 with $j = 1, 2, \cdots, \overline{M}$.

Consequently, we can derive the probability $\xi_{1 \to \overline{M}}^{MPR}$ that there are no slots with $1 \leq j \leq \overline{M}$ packets as follows:

$$\xi_{1 \to \overline{M}}^{MPR} = e^{-(\rho_1 + \rho_2 + \cdots + \rho_{\overline{M}})}. \tag{2.37}$$

Furthermore, since the event that all $\hat{h}$ packets fail to be received has two probabilities, i.e., (1) there are no slots with $1 \leq j \leq \overline{M}$ packets in the whole frame, and (2) there exists slots with $1 \leq j \leq \overline{M}$ packets, but all of these packets are unsuccessful. As a result, it holds that $\xi_{\hat{h}0}^{MPR} \geq \xi_{1 \to \overline{M}}^{MPR}$.

We thus can get the following inequalities:

$$\xi_{\hat{h}k}^{MPR} \leq 1 - \xi_{\hat{h}0}^{MPR} \leq 1 - e^{-(\rho_1 + \rho_2 + \cdots + \rho_{\overline{M}})} \leq 1 - e^{-\overline{M}\rho_{\overline{M}}}, k \geq 1, \tag{2.38}$$

where we use the fact that the probability of exact $k \geq 1$ successfully received packets among $\hat{h}$ packets is less than that of at least one packet received successfully in the first inequality. And the third inequality above follows from the monotonicity of ρ_j when $L = o(\hat{h}^{1-\epsilon_1})^*$, i.e.,

$$\rho_{\overline{M}} > \rho_{\overline{M}-1} > \cdots > \rho_2 > \rho_1.$$

In addition, we can also derive the following results:

$$0 \leq \lim_{\hat{h} \to \infty} \hat{h}^4 (1 - e^{-\overline{M}\rho_{\overline{M}}}) \leq \lim_{\hat{h} \to \infty} \frac{e^{\overline{M}\rho_{\overline{M}}} - 1}{(1/\hat{h}^4)} \leq \lim_{\hat{h} \to \infty} \frac{\epsilon_1 \hat{h}^{\overline{M}\epsilon_1 + 5}}{4\overline{M}! e^{\hat{h}^{\epsilon_1}}}$$

$$\leq \lim_{\hat{h} \to \infty} \frac{\prod_{x=0}^{\overline{M}-1+\lceil \frac{5}{\epsilon_1} \rceil} (\overline{M} + \frac{5}{\epsilon_1} - x)}{4\overline{M}! e^{\hat{h}^{\epsilon_1}}} \leq 0,$$

which means $1 - e^{-\overline{M}\rho_{\overline{M}}} \leq \frac{1}{\hat{h}^4}$. Using this inequality and recall (2.29), we have

$$\sum_{s=0}^{\infty} y_s P_{is} = \sum_{s=0}^{i-L} y_s P_{is} + \sum_{s=i-L+1}^{i} y_s P_{is} + \sum_{s=i+1}^{\infty} y_s P_{is}$$

$$\leq y_{i+1} + \sum_{s=i-L+1}^{i} y_s \sum_{n=0}^{s} \lambda_t(n) \xi_{\hat{h},n+i-s}$$

$$\leq \sum_{s=i-L+1}^{i} y_s \left(\sum_{n=0}^{L^{\frac{2}{3}}} \lambda_n + \sum_{n=L^{\frac{2}{3}}+1}^{s} \lambda_t(n) \xi_{\hat{h},n+i-s} \right) + y_{i+1}$$

$$\leq \sum_{s=i-L+1}^{i} y_s \left(\frac{1}{a_1^{L^{2/3}}} + \frac{1}{\hat{h}^4} \right) + y_{i+1}$$

$$\leq \frac{L}{(i - L + 2)^\theta} \left(\frac{1}{a_1^{L^{2/3}}} + \frac{1}{\hat{h}^4} \right) + \frac{1}{(i + 2)^\theta}$$

$$\leq 2(\beta i)^{-3} + \frac{1}{(i + 2)^\theta} \leq \frac{1}{(i + 1)^\theta}, \quad \text{as } i \to \infty. \tag{2.39}$$

Thus, according to Lemma 2.6, the backlog Markov chain $(X_t)_{t \geq 0}$ is transient when $L_t = o(\hat{h}^{1 - \epsilon_1})^*$.

Case 2: $L_t = \Theta(\hat{h})$ **and** $\Lambda > \alpha$
In this case, we have $\mathfrak{N}_t = L_t \Lambda > \hat{h}$. Using similar reasoning as (2.34), we have

$$\sum_{s=0}^{\infty} y_s P_{is} \leq \frac{\beta i + 1}{a_3^{\beta i}} + \frac{1}{(i + 2)^\theta} \leq \frac{1}{(i + 1)^\theta}, \quad \text{as } i \to \infty.$$

Therefore, $(X_t)_{t \geq 0}$ is also transient in this case and the proof of Theorem 2.4 is completed. $\qquad \square$

2.7 Discussion

In the sections above, we prove that the stability of FSA relies on the relationship between the frame size and the number of packets to be transmitted in this frame. In order to stabilize FSA systems, all the users should know the value of transmitted packets in the current frame. The state information, However, is imperfect in some scenarios such that the users do not have access to the value of packets to be transmitted in the frame. Fortunately, we can get its approximate value.

Recall (2.9), because N_t follows the Poisson distribution and Z_t follows the binomial distribution which can be approximated as the Poisson distribution, Y_t can also be approximated as a Poisson distributed random variable. According to Lemma 2.4, the value of Y_t sharply concentrates around its expectation, we thus use the following $E[Y_t]$ to approximate $\hat{h}$:

$$E[Y_t] = ip + (1 - p)E[N_{t-1} | X_t = i]$$

$$= ip + (1 - p) \sum_{w=0}^{i} \frac{w e^{-L_{t-1}\Lambda}(L_{t-1}\Lambda)^w}{w!}. \tag{2.40}$$

As a result, we can set the frame size following the control algorithm as follows:

$$L_t = c_1 \left(ip + (1 - p) \sum_{w=0}^{i} \frac{w e^{-L_{t-1}\Lambda}(L_{t-1}\Lambda)^w}{w!} \right), \tag{2.41}$$

$$L_0 = c_1 \vartheta, \tag{2.42}$$

where $X_0 = \vartheta$ means the initial number of packets in the system, and $c_1 = 1$ for FSA-SPR and $c_1 = \frac{1}{\alpha^*}$ for FSA-MPR.

By the above control algorithm, the frame size L_t only depends on the value of backlog population size i, so the original problem is translated to estimate the number of backlogs X_t, i.e., i. Fortunately, there exist several estimation approaches which exploit the channel feedback, such as the probability of a idle or collision slot, and the number of idle or collision slots.

According to the requirement on the estimation accuracy, we can select a rough estimator or an accurate estimator. Since $ip \leq L_t \leq i$ in (2.41), we can estimate X_t roughly so that the estimate $\tilde{X}_t = \Theta(X_t)$ in very short time, more specifically, in $\log(X_t)$ or $\log\log(X_t)$ slots [35]. While if the accurate result is required, we can use the additive estimator as in [7] and Kalman filter-based estimator as in our another work [36] to estimate the value of X_t and update L_t.

2.8 Numerical Results

In this section, we conduct simulations via MATLAB to verify our theoretical results by illustrating the evolution of the number of backlogs in each frame under different parameters with the following default settings: the initial number of backlogs $X_0 = 10^4$, the simulation duration $t_{\max} = 100$, $o(\hat{h}) \leq 0.01\hat{h}$, $O(\hat{h}) \geq 100\hat{h}$, $\alpha = 1$ in FSA-SPR and $\alpha = \alpha^* \in (\frac{\overline{M}-1}{e}, \overline{M})$ in FSA-MPR when $L_t = \Theta(\hat{h})$. To simulate FSA, each user first generates a random number among $[0, L_t - 1]$ uniformly and responds in the corresponding slot. And all results are obtained by taking the average of 100 trials.

2.8.1 Stability Properties of FSA

FSA-SPR Systems We start by investigating numerically the stability properties of FSA-SPR. As stated in Theorem 2.1, if $\Lambda < \frac{1}{e}$, the system is stable and unstable otherwise when $\alpha = 1$, we thus set the expected arrival rate per slot Λ to 0.3 and 0.37 for the analysis of stability and instability for $L_t = \hat{h}$, respectively. Moreover, we set a small $\Lambda = 0.01$ to analyze the instability for the cases $L_t = o(\hat{h})$ and $L_t = O(\hat{h})$.

As shown in Fig. 2.1a, for the case $L_t = \hat{h}$, the number of backlogs decreases to zero at a rate in proportion to the retransmission probability if $\Lambda < \frac{1}{e}$, while increasing gradually otherwise. This is due to the nature of FSA that frame size varies with the number of the sent packets to maximize the throughput per slot. Moreover, Fig. 2.1b, c illustrate the instability when $L_t = o(\hat{h})$ and $L_t = O(\hat{h})$. The numerical results is in accordance with the analytical results on FSA-SPR in Theorem 2.1.

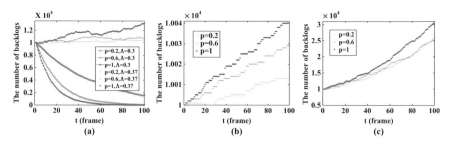

Fig. 2.1 The evolution of backlog population in FSA-SPR. (**a**) $L = \hat{h}$. (**b**) $L = o(h)$ and $\Lambda = 0.01$. (**c**) $L = O(h)$ and $\Lambda = 0.01$

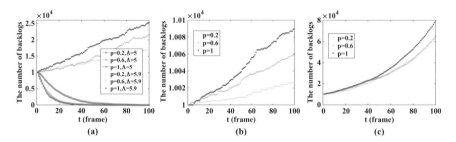

Fig. 2.2 The evolution of backlog population in FSA-MPR. (**a**) $L = \hat{h}/\alpha^*$. (**b**) $L = o(h)$ and $\Lambda = 0.01$. (**c**) $L = O(h)$ and $\Lambda = 0.01$

FSA-MPR System We then move to the FSA-MPR exploiting MPR model as in [6] with $\overline{M} = 10$. Recall Lemma 2.7, it can be derived that $\alpha^* = 10/1.37$ and the maximum stability region is $\hat{R}_1 = 5.814$. We thus set $\Lambda = 5$ and $\Lambda = 5.9$ for the stability analysis in the case $L_t = \hat{h}/\alpha^*$. From Fig. 2.2, we can see that the numerical results is in accordance with the analytical results on FSA-MPR.

2.8.2 Comparison Under Different Frame Sizes

Here we evaluate the performance difference when the frame size deviates from its optimum value that $\alpha = \frac{\hat{h}}{L_t} = 1$ in FSA-SPR and $\alpha = \alpha^*$ in FSA-MPR. To that end, we set $\Lambda = 0.3$ for FSA-SPR and $\Lambda = 5$ for FSA-MPR. As shown in Figs. 2.3 and 2.4, the performance degrades significantly when the frame size is not optimal because the throughput is reduced in this case.

Fig. 2.3 SPR: # of backlogs

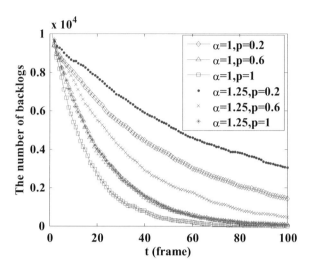

Fig. 2.4 MPR: # of backlogs

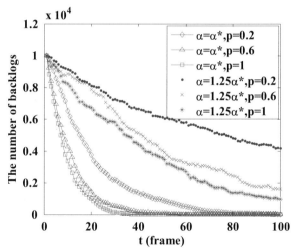

2.8.3 Comparison Between FSA-SPR and FSA-MPR

We further compare the performance of FSA-SPR and FSA-MPR. To that end, for both FSA-SPR and FSA-MPR, we set $\Lambda = 0.3$ and $L_t = \hat{h}$ where $\hat{h}$ is the number of backlogs in FSA-SPR maximizing the throughput of FSA-SPR. Although this setting is not optimal to FSA-MPR, we can also see from Fig. 2.5 that FSA-MPR remarkably outperforms FSA-SPR.

Fig. 2.5 SPR vs. MPR

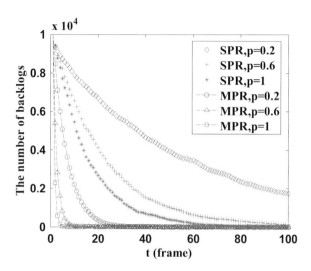

2.9 Conclusion

In this chapter, we have studied the stability of FSA-SPR and FSA-MPR by modeling the system backlog as a Markov chain. By employing drift analysis, we have obtained the closed-form conditions for the stability of FSA and shown that the stability region is maximised when the frame length equals the number of sent packets in FSA-SPR and the upper bound of stability region is maximised when the ratio of the number of sent packets to frame length equals in an order of magnitude the maximum multipacket reception capacity in FSA-MPR. Furthermore, to characterise system behavior in the instable region, we have mathematically demonstrated the existence of transience of the Markov chain. In addition, we conduct the numerical analysis to verify the theoretical results. Our results provide theoretical guidelines on the design of stable FSA-based protocols in practical applications such as RFID systems and M2M networks.

2.10 Proofs

2.10.1 Proof of Lemma 2.3

Proof We take into account two cases depending on whether r is finite.

Case 1: $\hat{r} = +\infty$

Recall the definition of the limit, fix $M > 0$ and pick a $w^* \geq 0$ such that $\lambda_{t-1}(w^*) \neq 0$, there exists an integer Q such that $r_{\hat{h}}^{SPR} > M$ for all $\hat{h} \geq Q$. Then fix such Q, we have

$$\sum_{w=0}^{i} \lambda_{t-1}(w) \sum_{h=0}^{i-w} B_{(i-w)}(h) r_{h+w}^{SPR} > \lambda_{t-1}(w^*) \sum_{h=0}^{i-w^*} B_{(i-w^*)}(h) r_{h+w^*}^{SPR}$$

$$> \lambda_{t-1}(w^*) \cdot \sum_{h=Q}^{i-w^*} B_{(i-w^*)}(h) r_{h+w^*}^{SPR} > M \lambda_{t-1}(w^*) \sum_{h=Q}^{i-w^*} B_{(i-w^*)}(h)$$

for all $i \geq Q + w^*$, since for any fixed Q and w^* it holds that

$$\lim_{i \to \infty} \sum_{h=Q}^{i-w^*} B_{(i-w^*)}(h) = 1. \tag{2.43}$$

Case 2: $\hat{r} < +\infty$

For $i > 2Q$, after some algebraic operations, we get

$$\sum_{w=0}^{i} \lambda_{t-1}(w) \sum_{h=0}^{i-w} B_{(i-w)}(h) r_{h+w}^{SPR}$$

$$= \sum_{w=0}^{Q} \lambda_{t-1}(w) \sum_{h=0}^{i-w} B_{(i-w)}(h) r_{h+w}^{SPR} + \sum_{w=Q+1}^{i} \lambda_{t-1}(w) \sum_{h=0}^{i-w} B_{(i-w)}(h) r_{h+w}^{SPR}$$

$$= \sum_{w=0}^{Q} \lambda_{t-1}(w) \sum_{h=0}^{2Q-w} B_{(i-w)}(h) r_{h+w}^{SPR} + \sum_{w=0}^{Q} \lambda_{t-1}(w) \sum_{h=2Q-w+1}^{i-w} B_{(i-w)}(h) r_{h+w}^{SPR}$$

$$+ \sum_{w=Q+1}^{i} \lambda_{t-1}(w) \sum_{h=0}^{i-w} B_{(i-w)}(h) r_{h+w}^{SPR}.$$

Now we proceed by calculate three items in the right hand side:

- In the first item, since $0 \leq r_{h+w} \leq h + w \leq 2Q$ and $0 < \lambda_{t-1}(w) < 1$ and (2.43) holds, we have $\sum_{w=0}^{Q} \lambda_{t-1}(w) \sum_{h=0}^{2Q-w} B_{(i-w)}(h) r_{h+w}^{SPR} \to 0$, as $i \to \infty$.
- In the second item, since it holds that $h + w > Q$ and (2.43), we have

$$\lim_{i \to \infty} \sum_{w=0}^{Q} \lambda_{t-1}(w) \sum_{h=2Q-w+1}^{i-w} B_{(i-w)}(h) r_{h+w}^{SPR}$$

$$= \lim_{i \to \infty} \hat{r} \sum_{w=0}^{Q} \lambda_{t-1}(w) \sum_{h=2Q-w+1}^{i-w} B_{(i-w)}(h) = \hat{r} \sum_{w=0}^{Q} \lambda_{t-1}(w).$$

- In the third item, since it holds that $h + w > Q$, we have

$$\lim_{i \to \infty} \sum_{w=Q+1}^{i} \lambda_{t-1}(w) \sum_{h=0}^{i-w} B_{(i-w)}(h) r_{h+w}^{SPR}$$

$$= \lim_{i \to \infty} \hat{r} \sum_{w=Q+1}^{i} \lambda_{t-1}(w) \sum_{h=0}^{i-w} B_{(i-w)}(h) = \hat{r} \sum_{w=Q+1}^{i} \lambda_{t-1}(w).$$

Consequently, we can get

$$\lim_{i \to \infty} \sum_{w=0}^{i} \lambda_{t-1}(w) \sum_{h=0}^{i-w} B_{(i-w)}(h) r_{h+w}^{SPR} = \hat{r} \sum_{w=0}^{Q} \lambda_{t-1}(w) + \hat{r} \sum_{w=Q+1}^{i} \lambda_{t-1}(w) = \hat{r},$$
(2.44)

which completes the proof of Case 2 and the lemma as well. $\qquad \square$

2.10.2 Proof of Lemma 2.6

Proof Recall the definition of transition probability, we can get the following equivalent:

$$\sum_{s} y_s P_{is} \le y_i \iff \sum_{s=1}^{\min(i,L_t)} (y_{i-s} - y_i) P_{i,i-s} + \sum_{s=1}^{\infty} (y_{i+s} - y_i) P_{i,i+s} \le 0.$$

With the following definitions:

$$\begin{cases} f'(i) = (i+1)^\theta \sum_{s=1}^{\min(i,L_t)} (\frac{1}{(i+1-s)^\theta} - \frac{1}{(i+1)^\theta}) \cdot \sum_{w=0}^{i} \\ \lambda_{t-1}(w) \sum_{h=\{s-w\}^+}^{i-w} B_{i-w}(h) \sum_{n=0}^{\min(L_t,h)-s} \lambda_t(n) \xi_{\hat{h},n+s}, \\ g'(i) = (i+1)^\theta \sum_{s=1}^{\infty} (\frac{1}{(i+1+s)^\theta} - \frac{1}{(i+1)^\theta}) \cdot \sum_{w=0}^{i} \\ \lambda_{t-1}(w) \sum_{h=0}^{i-w} B_{i-w}(h) \sum_{n=0}^{\min(\hat{h},L_t)} \lambda_t(n+s) \xi_{\hat{h}n}, \end{cases}$$
(2.45)

we have $\sum_s y_s P_{is} \le y_i \iff f'(i) + g'(i) \le 0$.

Moreover, the drift can be rewritten as

$$D_i = - \sum_{s=1}^{\min(i,L_t)} s P_{i,i-s} + \sum_{s=1}^{\infty} s P_{i,i+s} = f(i) + g(i),$$

where $f(h)$ and $g(h)$ are defined as

$$
\begin{cases}
f(i) = \sum_{s=1}^{\min\{i,L_t\}} s \sum_{w=0}^{i} \lambda_{t-1}(w) \sum_{h=\{s-w\}^+}^{i-w} \\
\quad B_{i-w}(h) \sum_{n=0}^{\min(L_t,\hat{h})-s} \lambda_t(n)\xi_{\hat{h},n+s}, \\
g(i) = \sum_{s=1}^{\infty} s \sum_{w=0}^{i} \lambda_{t-1}(w) \sum_{h=0}^{i-w} B_{i-w}(h) \\
\quad \sum_{n=0}^{\min(\hat{h},L_t)} \lambda_t(n+s)\xi_{\hat{h}n}.
\end{cases}
\tag{2.46}
$$

Therefore, $(X_t)_{t\geq 0}$ is transient if it holds that

$$
\lim_{i\to\infty}[f'(i) + g'(i) + \theta D_i] = 0.
\tag{2.47}
$$

Noticing that $D_i = f(i) + g(i)$, we prove (2.47) by showing that: (1) $\lim_{i\to\infty}[f'(i) + \theta f(i)] = 0$; (2) $\lim_{i\to\infty}[g'(i) + \theta g(i)] = 0$.
We first prove $\lim_{i\to\infty}[f'(i) + \theta f(i)] = 0$.
From (2.45) and (2.46), we get

$$
f'(i) + \theta f(i) = (i+1) \sum_{s=1}^{\min(i,L_t)} \left[\left(\frac{i+1}{i+1-s}\right)^\theta - 1 - \frac{\theta s}{i+1}\right]
$$

$$
\cdot \sum_{w=0}^{i} \lambda_{t-1}(w) \sum_{h=\{s-w\}^+}^{i-w} B_{i-w}(h) \sum_{n=0}^{L_t-s} \lambda_t(n)\xi_{\hat{h},n+s},
\tag{2.48}
$$

which is nonnegative since

$$
\left(\frac{i+1}{i+1-s}\right)^\theta - 1 - \frac{\theta s}{i+1} > 0, \ \forall 1 \leq s \leq i.
$$

and thus

$$
f'(i) + \theta f(i) \leq (i+1) \sum_{s=1}^{L_t} \left[\left(\frac{i+1}{i+1-s}\right)^\theta - 1 - \frac{\theta s}{i+1}\right].
$$

Given $0 < v \leq L_t < i$, define $m_i(v)$ as

$$
m_i(v) = \frac{i+1}{v^2}\left[\left(\frac{i+1}{i+1-v}\right)^\theta - 1\right] - \frac{\theta}{v}
$$

which is positive and nondecreasing in v for $i \geq 1$, we get

$$
m_i\left(\left\lfloor\frac{i+1}{2}\right\rfloor\right) \leq \frac{1}{i+1}[4(2^\theta - 1) - 2\theta] \triangleq \frac{A}{i+1},
$$

where A is a positive constant only depending on θ. Thus,

$$f'(i) + \theta f(i) \le \sum_{s=1}^{L_t} s^2 m_i(s) \le \frac{A}{i+1} \sum_{s=1}^{L_t} s^2 \le \frac{AL_t(L_t+1)(2L_t+1)}{6(i+1)}.$$

Since $\lim_{i \to \infty} \frac{AL_t(L_t+1)(2L_t+1)}{6(i+1)} = 0$ when $L_t = o((\ln i)^4)^*$, it thus holds that $\lim_{i \to \infty}[f'(i) + \theta f(i)] = 0$.

We then prove $\lim_{i \to \infty}[g'(i) + \theta g(i)] = 0$.

From (2.45) and (2.46), we get

$$g'(i) + \theta g(i) = (i+1) \sum_{s=1}^{\infty} \left[\left(\frac{i+1}{i+1+s}\right)^{\theta} - 1 + \frac{\theta s}{i+1} \right]$$

$$\cdot \sum_{w=0}^{i} \lambda_{t-1}(w) \sum_{h=0}^{i-w} B_{i-w}(h) \sum_{n=0}^{\min(\hat{h},L_t)} \lambda_t(n+s)\xi_{\hat{h}n}.$$

Since $\left[\left(\frac{i+1}{i+1+s}\right)^{\theta} - 1 + \frac{\theta s}{i+1} \right] \ge 0$, after some algebraic operations, we have

$$g'(i) + \theta g(i) \le (i+1) \sum_{n=1}^{\infty} \lambda_t(n) \sum_{s=1}^{n} \left[\left(\frac{i+1}{i+1+s}\right)^{\theta} - 1 + \frac{\theta s}{i+1} \right] \cdot \xi_{\hat{h},n-s}$$

Using the following inequalities for $v \ge 0$ and $0 < \theta < 1$,

$$0 \le \frac{1}{(1+v)^{\theta}} - 1 + \theta v \le \theta(1+\theta)\frac{v^2}{2}, \tag{2.49}$$

we have

$$0 \le g'(i) + \theta g(i)$$

$$\le \frac{\theta(\theta+1)}{2}(i+1) \sum_{w=0}^{i} \lambda_{t-1}(w) \sum_{h=0}^{i-w} B_{i-w}(h) \sum_{n=1}^{N} \lambda_t(n) \sum_{s=1}^{n} \frac{s^2}{(i+1)^2}\xi_{\hat{h},n-s}$$

$$+ \theta(i+1) \sum_{w=0}^{i} \lambda_{t-1}(w) \sum_{h=0}^{i-w} B_{i-w}(h) \sum_{n=N+1}^{\infty} \lambda_t(n) \sum_{s=1}^{n} \frac{s}{i+1}\xi_{\hat{h},n-s}$$

$$\le \frac{1}{2(i+1)} \sum_{n=1}^{N} n^2 \lambda_t(n) + \sum_{n=N+1}^{\infty} n\lambda_t(n).$$

When $L_t = o((lni)^4)^*$, according to (2.23) in Lemma 2.4, we can choose $N = i^{1/3} \gg L_t \Lambda$ such that

$$P\{N_t \geq i^{1/3}\} \leq \frac{e^{-\lambda}(e\lambda)^{i^{1/3}}}{(i^{1/3})^{i^{1/3}}} \leq e^{-i^{1/3}(\frac{L_t\Lambda}{i^{1/3}}-1)} \left(\frac{L_t\Lambda}{i^{1/3}}\right)^{i^{1/3}} \leq \frac{1}{a_4^{i^{1/3}}}, \qquad (2.50)$$

where $a_4 \triangleq \frac{i^{1/3}}{\Lambda L_t} e^{\frac{\Lambda L_t}{i^{1/3}}-1} > \frac{1}{2}i^{1/3}$.

Then, fix $N = i^{1/3}$, we have

$$\frac{1}{2(i+1)} \sum_{n=1}^{N} n^2 \lambda_t(n) \leq \frac{i^{1/3}}{2(i+1)} L_t\Lambda \leq i^{-1/3}. \qquad (2.51)$$

As a consequence, we have $\lim_{i\to\infty}[g'(i) + \theta g(i)] = 0$ at last, which completes the proof of the second part and also Lemma 2.6.   □

2.10.3 Proof of Lemma 2.7

Proof Recall $\hat{R}_1$, we have $\hat{R}_1 \leq \sum_{x_0=1}^{\overline{M}} x_0 e^{-\alpha} \frac{\alpha^{x_0}}{x_0!} \triangleq \Phi(\alpha)$. We then rewrite the upper bound $\Phi(\alpha)$ of $\hat{R}_1$ as

$$\Phi(\alpha) = e^{-\alpha} \sum_{m=1}^{\overline{M}} \frac{\alpha^m}{(m-1)!},$$

whose derivative can be calculated as

$$\Phi'(\alpha) = e^{-\alpha} \left[\sum_{m=0}^{\overline{M}-1} \frac{\alpha^m}{m!} - \frac{\alpha^{\overline{M}}}{(\overline{M}-1)!} \right]. \qquad (2.52)$$

We distinguish two cases to look for α^*.

Case 1: $\alpha \geq \overline{M}$

Since it holds that $N! \leq N^{N-1}$ for $\forall N \in \mathbb{N}$, we can get

$$\Phi'(\alpha) < \frac{e^{-\alpha}}{(\overline{M}-1)!} \cdot \left(\sum_{m=1}^{\overline{M}} \overline{M}^{\overline{M}-m} \alpha^{m-1} - \alpha^{\overline{M}} \right) < \frac{e^{-\alpha}}{(\overline{M}-1)!} \cdot \left(\overline{M}\alpha^{\overline{M}-1} - \alpha^{\overline{M}}\overline{M} \right) < 0,$$

meaning that $\Phi(\alpha)$ monotonously decreases when $\alpha \geq \overline{M}$.

Case 2: $\alpha \leq \frac{\overline{M}-1}{e}$

Substituting the inequality $N! \geq (\frac{N}{e})^N$ into (2.52) yields

$$\Phi'(\alpha) \geq \frac{e^{-\alpha}}{(\overline{M}-1)!} \cdot \left[(\frac{\overline{M}-1}{e})^{\overline{M}-1} + \alpha(\frac{\overline{M}-1}{e})^{\overline{M}-1} - \alpha^{\overline{M}}\right]$$

$$\geq \frac{e^{-\alpha}}{(\overline{M}-1)!} \cdot \left(\alpha^{\overline{M}-1} + \alpha^{\overline{M}} - \alpha^{\overline{M}}\right) > 0,$$

meaning that $\Phi(\alpha)$ monotonously increases as $\alpha \leq \frac{\overline{M}-1}{e}$.

Combining the analysis in both cases, we have proved that α^* maximising $\Phi(\alpha)$ falls into the interval $(\frac{\overline{M}-1}{e}, \overline{M})$, i.e., $\alpha^* = \Theta(\overline{M})$. $\qquad\square$

References

1. N. Abramson, The Aloha system: another alternative for computer communications, in *Proceedings of Fall 1970 AFIPS Fall Joint Computer Conference* (1970), pp. 281–285
2. L.G. Robert, ALOHA packet system with and without slots and capture. ACM SIGCOMM Comput. Commun. Rev. **5**(2), 28–42 (1975)
3. H. Okada, Y. Igarashi, Y. Nakanishi, Analysis and application of framed ALOHA channel in satellite packet switching networks-fadra method. Electron. Commun. Jpn. **60**, 72–80 (1977)
4. H.-J. Noh, J.-K. Lee, J.-S. Lim, Anc-aloha: analog network coding ALOHA for satellite networks. IEEE Commun. Lett. **18**(6), 957–960 (2014)
5. S. Vasudevan, D. Towsley, D. Goeckel, R. Khalili, Neighbor discovery in wireless networks and the coupon collector's problem, in *ACM MobiCom* (2009), pp. 181–192
6. W. Zeng, S. Vasudevan, X. Chen, B. Wang, A. Russell, W. Wei, Neighbor discovery in wireless networks with multipacket reception, in *ACM MobiHoc* (2011), p. 3
7. H. Wu, C. Zhu, R.J. La, X. Liu, Fasa: accelerated S-ALOHA using access history for event-driven M2M communications. IEEE/ACM Trans. Netw. **21**(6), 1904–1917 (2013)
8. F. Vázquez Gallego, J. Alonso-Zarate, L. Alonso, Energy and delay analysis of contention resolution mechanisms for machine-to-machine networks based on low-power WiFi, in *IEEE ICC* (2013), pp. 2235–2240
9. Y. Zhu, W. Jiang, Q. Zhang, H. Guan, Energy-efficient identification in large-scale RFID systems with handheld reader. IEEE Trans. Parallel Distrib. Syst. **25**(5), 1211–1222 (2014)
10. X. Liu, K. Li, G. Min, K. Lin, B. Xiao, Y. Shen, W. Qu, Efficient unknown tag identification protocols in large-scale RFID systems. IEEE Trans. Parallel Distrib. Syst. **25**(12), 3145–3155 (2014)
11. EPCglobal Inc., Radio-frequency identity protocols class-1 generation-2 UHF RFID protocol for communications at 860 MHz–960 MHz version 1.0.9, EPCglobal Inc., vol. 17 (2005)
12. C. Bordenave, D. McDonald, A. Proutiere, Asymptotic stability region of slotted ALOHA. IEEE Trans. Inf. Theory **58**(9), 5841–5855 (2012)
13. S.C. Kompalli, R.R. Mazumdar, On the stability of finite queue slotted aloha protocol. IEEE Trans. Inf. Theory **59**(10), 6357–6366 (2013)
14. F. Farhadi, F. Ashtiani, Stability region of a slotted aloha network with k-exponential backoff. arXiv preprint:1406.4448 (2014)
15. N. Johnson, S. Kotz, *Urn Models and Their Application: An Approach to Modern Discrete Probability Theory* (Wiley, New York, 1977)

16. B.S. Tsybakov, V.A. Mikhailov, Ergodicity of a slotted ALOHA system. Probl. Peredachi Inf. **15**(4), 73–87 (1979)
17. R.R. Rao, A. Ephremides, On the stability of interacting queues in a multiple-access system. IEEE Trans. Inf. Theory **34**(5), 918–930 (1988)
18. W. Szpankowski, Stability conditions for some distributed systems: buffered random access systems. Adv. Appl. Probab. **26**(2), 498–515 (1994)
19. S. Ghez, S. Verdu, S.C. Schwartz, Stability properties of slotted Aloha with multipacket reception capability. IEEE Trans. Autom. Control **33**(7), 640–649 (1988)
20. S. Ghez, S. Verdu, S.C. Schwartz, Optimal decentralized control in the random access multipacket channel. IEEE Trans. Autom. Control **34**(11), 1153–1163 (1989)
21. J. Sant, V. Sharma, Performance analysis of a slotted-ALOHA protocol on a capture channel with fading. Queueing Syst. **34**(1–4), 1–35 (2000)
22. V. Naware, G. Mergen, L. Tong, Stability and delay of finite-user slotted ALOHA with multipacket reception. IEEE Trans. Inf. Theory **51**(7), 2636–2656 (2005)
23. H. Inaltekin, M. Chiang, H.V. Poor, S.B. Wicker, Selfish random access over wireless channels with multipacket reception. IEEE J. Sel. Areas Commun. **30**(1), 138–152 (2012)
24. J. Jeon, A. Ephremides, On the stability of random multiple access with stochastic energy harvesting. IEEE J. Sel. Areas Commun. **33**(3), 571–584 (2015)
25. J.E. Wieselthier, A. Ephremides, A. Larry, An exact analysis and performance evaluation of framed ALOHA with capture. IEEE Trans. Commun. **37**(2), 125–137 (1989)
26. F.C. Schoute, Dynamic frame length ALOHA. IEEE Trans. Commun. **31**(4), 565–568 (1983)
27. Z.G. Prodanoff, Optimal frame size analysis for framed slotted AlOHA based RFID networks. Comput. Commun. **33**(5), 648–653 (2010)
28. L. Barletta, F. Borgonovo, M. Cesana, A formal proof of the optimal frame setting for dynamic-frame aloha with known population size. IEEE Trans. Inf. Theory **60**(11), 7221–7230 (2014)
29. H. Vogt, Efficient object identification with passive RFID tags, in *International Conference on Pervasive Computing* (2002), pp. 98–113
30. A.G. Pakes, Some conditions for ergodicity and recurrence of Markov chains. Oper. Res. **17**(6), 1058–1061 (1969)
31. M. Kaplan, A sufficient condition for nonergodicity of a Markov chain. IEEE Trans. Inf. Theory **25**(4), 470–471 (1979)
32. M. Mitzenmacher, E. Upfal, *Probability and Computing: Randomized Algorithms and Probabilistic Analysis* (Cambridge University Press, Cambridge, 2005)
33. J.F. Mertens, E.S. Cahn, S. Zamir, Necessary and sufficient conditions for recurrence and transience of Markov chains, in terms of inequalities. J. Appl. Probab. **15**(4), 848–851 (1978)
34. W. Feller, *An Introduction to Probability Theory and its Applications*, vol. 2 (Wiley, Hoboken, 2008)
35. B. Chen, Z. Zhou, H. Yu, Understanding RFID counting protocols, in *ACM MobiHoc* (2013), pp. 291–302
36. J. Yu, L. Chen, From static to dynamic tag population estimation: an extended Kalman Filter perspective. arXiv preprint:1511.08355 (2015)

Chapter 3
From Static to Dynamic Tag Population Estimation: An Extended Kalman Filter Perspective

Chapter Roadmap The rest of this chapter is organised as follows. Section 3.1 explains the motivation of studying dynamic tag population estimation and summarizes the contributions. Section 3.2 gives a brief review of existing tag population estimation algorithms. Section 3.3 introduces background knowledge on Extended Kalman filter and stochastic process. Section 3.4 formulates the dynamic tag estimation problem. Sections 3.5 and 3.6 present tag estimation in static and dynamic systems. Section 3.7 conducts theoretical performance analysis. Section 3.8 discusses the issues in practical environment. Section 3.9 shows the experimental results. Section 3.10 gives the summary.

3.1 Introduction

3.1.1 Context and Motivation

Recent years have witnessed an unprecedented development and application of the radio frequency identification (RFID) technology. As a promising low-cost technology, RFID is widely utilized in various applications ranging from inventory control [1, 2], supply chain management [3] to tracking/location [4, 5]. A standard RFID system has two types of devices: a set of RFID tags and one or multiple RFID readers (simply called *tags* and *readers*). A tag is typically a low-cost microchip labeled with a unique serial number (ID) to identify an object. A reader, on the other hand, is equipped with an antenna and can collect the information of tags within its coverage area.

Tag population estimation and counting is a fundamental functionality for many RFID applications such as warehouse management, inventory control and tag identification. For example, quickly and accurately estimating the number of tagged objects is crucial in establishing inventory reports for large retailers such

© Springer International Publishing AG, part of Springer Nature 2019
J. Yu, L. Chen, *Tag Counting and Monitoring in Large-Scale RFID Systems*,
https://doi.org/10.1007/978-3-319-91992-8_3

as Wal-Mart [6]. Due to the paramount practical importance of tag population estimation, a large body of studies [7–11] have been devoted to the design of efficient estimation algorithms. Most of them, as reviewed in Sect. 3.2, are focused on the static scenario where the tag population is constant during the estimation process. However, many practical RFID applications, such as logistic control, are dynamic in the sense that tags may be activated or terminated as specialized in C1G2 standard [12], or the tagged objects may enter and/or leave the reader's covered area frequently, thus resulting in tag population variation. In such dynamic applications, a fundamental research question is how to design efficient algorithms to dynamically trace the tag population quickly and accurately.

3.1.2 Summary of Contributions

In this chapter, we develop a generic framework of stable and accurate tag population estimation schemes for both static and dynamic RFID systems. By generic, we mean that our framework both supports the real-time monitoring and can estimate the number of tags accurately without any prior knowledge on the tag arrival and departure patterns. Our design is based on the extended Kalman filter (EKF) [13], a powerful tool in optimal estimation and system control. By performing Lyapunov drift analysis, we mathematically prove the efficiency and stability of our framework in terms of the boundedness of estimation error.

The main technical contributions of this chapter are articulated as follows. We formulate the system dynamics of the tag population for both static and dynamic RFID systems where the number of tags remains constant and varies during the estimation process. We design an EKF-based population estimation algorithm for static RFID systems and further enhance it to dynamic RFID systems by leveraging the cumulative sum control chart (CUSUM) to detect the population change. By employing Lyapunov drift analysis, we mathematically characterise the performance of the proposed framework in terms of estimation accuracy and convergence speed by deriving the closed-form conditions on the design parameters under which our scheme can stabilise around the real population size with bounded relative estimation error that tends to zero within exponential convergence rate. To the best of our knowledge, our work is the first theoretical framework that dynamically traces the tag population with closed form conditions on the estimation stability and accuracy.

3.2 Related Work

Due to its fundamental importance, tag population estimation has received significant research attention, which we briefly review in this section.

3.2.1   *Tag Population Estimation for Static RFID Systems*

Most of existing works are focused on the static scenario where the tag population is constant during the estimation process. The central question there is to design efficient algorithms quickly and accurately estimating the static tag population. Kodialam et al. design an estimator called PZE which uses the probabilistic properties of empty and collision slots to estimate the tag population size [14]. The authors then further enhance PZE by taking the average of the probability of idle slots in multiple frames as an estimator in order to eliminate the constant additive bias [7]. Han et al. exploit the average number of idle slots before the first non-empty slots to estimate the tag population size [15]. Later, Qian et al. develop Lottery-Frame scheme that employs geometrically distributed hash function such that the jth slot is chosen with prob. $\frac{1}{2^{j+1}}$ [9]. As a result, the first idle slot approaches around the logarithm of the tag population and the frame size can be reduced to the logarithm of the tag population, thus reducing the estimation time. Subsequently, a new estimation scheme called ART is proposed in [10] based on the average length of consecutive non-empty slots. The design rational of ART is that the average length of consecutive non-empty slots is correlated to the tag population. ART is shown to have smaller variance than prior schemes. More recently, Zheng et al. propose another estimation algorithm, ZOE, where each frame just has a single slot and the random variable indicating whether a slot is idle follows Bernoulli distribution [11]. The average of multiple individual observations is used to estimate the tag population.

We would like to point out that the above research work does not consider the estimation problem for dynamic RFID systems and thus may fail to monitor the system dynamics in real time. Specifically, in typical static tag population estimation schemes, the final estimation result is the average of the outputs of multi-round executions. When applied to dynamic tag population estimation, additional estimation error occurs due to the variation of the tag population size during the estimation process.

3.2.2   *Tag Population Estimation for Dynamic RFID Systems*

Only a few propositions have tackled the dynamic scenario. The works in [16] and [17] consider specific tag mobility patterns that the tags move along the conveyor in a constant speed, while tags may move in and out by different workers from different positions, so these two algorithm cannot be applicable to generic dynamic scenarios. Subsequently, Xiao et al. develop a differential estimation algorithm, ZDE, in dynamic RFID systems to estimate the number of arriving and removed tags [18]. More recently, they further generalize ZDE by taking into account the snapshots of variable frame sizes [19]. Though the algorithms in [18] and [19] can monitor the dynamic RFID systems, they may fail to estimate the tag population size accurately, because they must use the same hash seed in the whole monitoring process, which cannot reduce the estimation variance. Using the same seed is an effective way in tracing tag departure and arrival, however, it may significantly limit the estimation accuracy, even in the static case.

Besides the limitations above, prior works do not provide formal analysis on the stability and the convergence rate. To fill this vide, we develop a generic framework for tag population estimation in dynamic RFID systems. By generic, we mean that our framework can both support real-time monitoring and estimate the number of tags accurately without the requirement for any prior knowledge on the tag arrival and departure patterns. As another distinguished feature, the efficiency and stability of our framework is mathematically established.

3.3 Technical Preliminaries

In this section, we briefly introduce the extended Kalman filter and some fundamental concepts and results in stochastic process which are useful in the subsequent analysis. The main notations used in this chapter are listed in Table 3.1.

Table 3.1 Main notations

z_k	System state in frame k: tag population
y_k	Measurement in frame k: idle slot frequency
$\hat{z}_{k+1|k}$	Priori prediction of z_{k+1}
$\hat{z}_{k|k}$	Posteriori estimate of z_k
$P_{k+1|k}$	Priori pseudo estimate covariance
$P_{k|k}$	Posteriori pseudo estimate covariance
v_k	Measurement residual in frame k
K_k	Kalman gain in frame k
Q_k, R_k	Two tunable parameters in frame k
$e_{k|k-1}$	Estimation error in frame k
Φ_k	Normalization of v_k
L_k	The length of frame k
$Rs_k, h(\cdot)$	Random seed in frame k and Hash function
r_k	Persistence probability in frame k
N_k	The number of idle slots in frame k
$p(z_k)$	Probability of an idle slot in frame k
$u_k, Var[u_k]$	Gaussian random variable and Variance of u_k
ϕ_k	Controllable parameter
w_k	Random variable: variation of tag population
θ, Υ_k	CUSUM threshold and reference value
ϵ	Upper bound of initial estimation error
λ_k, δ_k	Upper bounds of $E[w_k]$ and $E[w_k^2]$

3.3.1 Extended Kalman Filter

The extended Kalman filter is a powerful tool to estimate system state in nonlinear discrete-time systems. Formally, a nonlinear discrete-time system can be described as follows:

$$z_{k+1} = f(z_k, x_k) + w_k^* \tag{3.1}$$

$$y_k = h(z_k) + u_k^*, \tag{3.2}$$

where $z_{k+1} \in \mathbb{R}^n$ denotes the state of the system, $x_k \in \mathbb{R}^d$ is the controlled inputs and $y_k \in \mathbb{R}^m$ stands for the measurement observed from the system. The uncorrelated stochastic variables $w_k^* \in \mathbb{R}^n$ and $u_k^* \in \mathbb{R}^m$ denote the process noise and the measurement noise, respectively. The functions f and h are assumed to be the continuously differentiable.

For the above system, we introduce an EKF-based state estimator given in Definition 3.1.

Definition 3.1 (Extended Kalman Filter [13]) A two-step discrete-time extended Kalman filter consists of state prediction and measurement update, defined as

1) Time update (prediction)

$$\hat{z}_{k+1|k} = f(\hat{z}_{k|k}, x_k) \tag{3.3}$$

$$P_{k+1|k} = P_{k|k} + Q_k, \tag{3.4}$$

2) Measurement update (correction)

$$\hat{z}_{k+1|k+1} = f(\hat{z}_{k+1|k}, x_k) + K_{k+1} v_{k+1} \tag{3.5}$$

$$P_{k+1|k+1} = P_{k+1|k} (1 - K_{k+1} C_{k+1}) \tag{3.6}$$

$$K_{k+1} = \frac{P_{k+1|k} C_{k+1}}{P_{k+1|k} C_{k+1}^2 + R_{k+1}}, \tag{3.7}$$

where

$$v_{k+1} = y_{k+1} - h(\hat{z}_{k+1|k}) \tag{3.8}$$

$$C_{k+1} = \frac{\partial h(z_{k+1})}{\partial z_{k+1}} \bigg|_{z_{k+1} = \hat{z}_{k+1|k}}. \tag{3.9}$$

Remark In the above definition of extended Kalman filter, the parameters can be interpreted in our context as follows:

- $\hat{z}_{k+1|k}$ is the prediction of z_{k+1} at the beginning of frame $k + 1$ given by the previous state estimate, while $\hat{z}_{k+1|k+1}$ is the estimate of z_{k+1} after the adjustment based on the measure at the end of frame $k + 1$.

- v_{k+1}, referred to as innovation, is the measurement residual in frame $k + 1$. It represents the estimated error of the measure.
- K_{k+1} is the Kalman gain. With reference to equation (3.5), it weighs the innovation v_{k+1} w.r.t. $f(\hat{z}_{k+1|k}, x_k)$.
- $P_{k+1|k}$ and $P_{k+1|k+1}$, in contrast to the linear case, are not equal to the covariance of estimation error of the system state. Here, we will refer to them as pseudo-covariance.
- Q_k and R_k are two tunable parameters which play the role as that of the covariance of the process and measurement noises in linear stochastic systems to achieve optimal filtering in the maximum likelihood sense. We will show later that Q_k and R_k also play an important role in improving the stability and convergence of our EKF-based estimators.

3.3.2 Boundedness of Stochastic Process

In order to analyse the stability of an estimation algorithm, we need to check the boundedness of the estimation error defined as follows:

$$e_{k|k-1} \triangleq z_k - \hat{z}_{k|k-1}. \tag{3.10}$$

Due to probabilistic nature of the estimation algorithm, the estimation process is a stochastic process. Thus, we further introduce the following two mathematical definitions [20, 21] on the boundedness of stochastic process.

Definition 3.2 (Boundedness of Random Variable) The stochastic process of the estimation error $e_{k|k-1}$ is said to be bounded with probability one (w.p.o.), if there exists $X > 0$ such that

$$\lim_{k \to \infty} \sup_{k \geq 1} \mathbb{P}\{|e_{k|k-1}| > X\} = 0. \tag{3.11}$$

Definition 3.3 (Boundedness in Mean Square) The stochastic process $e_{k|k-1}$ is said to be exponentially bounded in the mean square with exponent ζ, if there exist real numbers $\psi_1, \psi_2 > 0$ and $0 < \zeta < 1$ such that

$$E[e_{k|k-1}^2] \leq \psi_1 e_{1|0}^2 \zeta^{k-1} + \psi_2. \tag{3.12}$$

To investigate the boundedness defined in Definition 3.2 and 3.3, we introduce the following lemma [22].

Lemma 3.1 *Given a stochastic process $V_k(e_{k|k-1})$ and constants $\underline{\beta}, \overline{\beta}, \tau > 0$ and $0 < \alpha \leq 1$ with the following properties:*

$$\underline{\beta} e_{k|k-1}^2 \leq V_k(e_{k|k-1}) \leq \overline{\beta} e_{k|k-1}^2, \tag{3.13}$$

$$E[V_{k+1}(e_{k+1|k})|e_{k|k-1}] - V_k(e_{k|k-1}) \leq -\alpha V_k(e_{k|k-1}) + \tau, \tag{3.14}$$

then for any $k \geq 1$ it holds that

- *the stochastic process $e_{k|k-1}$ is exponentially bounded in the mean square, i.e.,*

$$E[e_{k|k-1}^2] \leq \frac{\overline{\beta}}{\underline{\beta}} E[e_{1|0}^2](1-\alpha)^{k-1} + \frac{\tau}{\underline{\beta}} \sum_{j=1}^{k-2} (1-\alpha)^j \leq \frac{\overline{\beta}}{\underline{\beta}} E[e_{1|0}^2](1-\alpha)^{k-1} + \frac{\tau}{\underline{\beta}\alpha},$$

(3.15)

- *the stochastic process $e_{k|k-1}$ is bounded w.p.o..*

From Lemma 3.1, it can be known that if we can construct $V_k(e_{k|k-1})$, a function of $e_{k|k-1}$, such that both its drift and $\frac{V_k(e_{k|k-1})}{e_{k|k-1}^2}$ are bounded, i.e, (3.14) and (3.13) hold, then $e_{k|k-1}$ is also bounded and the convergence rate depends on constant α mostly. Besides, it can be noted that Lemma 3.1 can only be implemented offline. To address this limit, we adjust Lemma 3.1 to an online version with time-varying parameters, which can be proven by the same method as in [21] and [23].

Lemma 3.2 *If there exist a stochastic process $V_k(e_{k|k-1})$ and real numbers β^*, β_k, $\tau_k > 0$ and $0 < \alpha_k^* \leq 1$ with the following properties:*

$$V_1(e_{1|0}) \leq \beta^* e_{1|0}^2,$$

(3.16)

$$\beta_k e_{k|k-1}^2 \leq V_k(e_{k|k-1}),$$

(3.17)

$$E[V_{k+1}(e_{k+1|k})|e_{k|k-1}] - V_k(e_{k|k-1}) \leq -\alpha_k^* V_k(e_{k|k-1}) + \tau_k;$$

(3.18)

then for any $k \geq 1$ it holds that

- *the stochastic process $e_{k|k-1}$ is exponentially bounded in the mean square, i.e.,*

$$E[e_{k|k-1}^2] \leq \frac{\beta^*}{\beta_k} E[e_{1|0}^2] \prod_{i=1}^{k-1} (1 - \alpha_i^*) + \frac{1}{\beta_k} \sum_{i=1}^{k-2} \tau_{k-i-1} \prod_{j=1}^{i} (1 - \alpha_{k-j}^*),$$

(3.19)

- *the stochastic process $e_{k|k-1}$ is bounded w.p.o..*

Remark The conditions in Lemma 3.2 can be interpreted as follows: To prove the boundedness of $e_{k|k-1}$, it is sufficient by constructing a function $V_k(e_{k|k-1})$ such that both its drift, i.e, (3.18), and $\frac{V_k(e_{k|k-1})}{e_{k|k-1}^2}$, i.e, (3.16), (3.17), are bounded.

3.4   System Model and Problem Formulation

3.4.1   System Model

Consider a RFID system consisting of a reader and a mass of tags operating on one frequency channel. The number of tags is unknown a priori and can be constant or dynamic (time-varying), which we refer to as *static* and *dynamic* systems, respectively throughout this chapter. The MAC protocol for the RFID system is the standard framed-slotted ALOHA protocol, where the standard *Listen-before-Talk* mechanism is employed by the tags to respond the reader's interrogation [24].

The reader initiates a series of frames indexed by an integer $k \in \mathbb{Z}_+$. Each individual frame, referred to as a round, consists of a number of slots. The reader starts frame k by broadcasting a `begin-round` command with frame size L_k, persistence probability r_k and a random seed Rs_k. When a tag receives a `begin-round` command, it uses a hash function $h(\cdot)$, L_k, Rs_k, and its ID to generate a random number i in the range $[0, L_k - 1]$ and reply in slot i of frame k with probability r_k.[1]

Since every tag picks its own response slot individually, there may be zero, one, or more than one tags transmitting in a slot, which are referred to as *idle*, *singleton*, and *collision* slots, respectively. The reader is not assumed to be able to distinguish between a singleton or a collision slot, but it can detect an idle slot. We term both singleton and collision slots as *occupied* slots throughout this chapter. By collecting all replies in a frame, the reader can generate a bit-string B_k illustrated as $B_k = \{\cdots |0|0|1|0|1|1| \cdots\}$, where '0' indicates an idle slot, and '1' stands for an occupied one.

Subsequently, the reader finalizes the current frame by sending an `end round` command. Based on the number of idle slots, i.e., the number of '0' in B_k, the reader runs the estimation algorithm, detailed in the next section, to trace the tag population.

3.4.2   Tag Population Estimation Problem

Our objective is to design a stable and accurate tag population estimation algorithm for both static and dynamic systems. By stable and accurate we mean that

- the estimation error of our algorithm is bounded in the sense of Definition 3.2 and 3.3 and the relative estimation error tends to zero;
- the estimated population size converges to the real value with exponential rate.

---

[1]The outputs of the hash function have a uniform distribution such that the tag can choose any slot within the round with the equal probability.

Mathematically, we consider a large-scale RFID system of a reader and a set of tags with the unknown size z_k in frame k which can be static or dynamic. Denote by $\hat{z}_{k|k-1}$ the prior estimate of z_k in the beginning of frame k. At the end of frame k, the reader updates the estimate $\hat{z}_{k|k-1}$ to $\hat{z}_{k|k}$ by running the estimation algorithm. Our designed estimation scheme need to guarantee the following properties:

- $\displaystyle \lim_{z_k \to \infty} \left| \frac{\hat{z}_{k|k-1} - z_k}{z_k} \right| = 0;$
- the converges rate is exponential.

3.5 Tag Population Estimation: Static Systems

In this section, we focus on the baseline scenario of static systems where the tag population is constant during the estimation process. We first establish the discrete-time model for the system dynamics and the measurement model using the bit string B_k observed during frame k. We then present our EKF-based estimation algorithm.

3.5.1 System Dynamics and Measurement Model

Consider the static RFID systems where the tag population stays constant, the system state evolves as

$$z_{k+1} = z_k, \tag{3.20}$$

meaning that the number of tags z_{k+1} in the system in frame $k + 1$ equals that in frame k.

In order to estimate z_k, we leverage the measurement on the number of idle slots during a frame. To start, we study the stochastic characteristics of the number of idle slots.

Assume that the initial tag population z_0 falls in the interval $z_0 \in [\underline{z}_0, \overline{z}_0]$, yet the exact value of z_0 is unknown and should be estimated. The range $[\underline{z}_0, \overline{z}_0]$ can be a very coarse estimation that can be obtained by any existing population estimation method. Recall the system model that in frame k, the reader probes the tags with the frame size L_k. Denote by variable N_k the number of idle slots in frame k, that is, the number of '0's in B_k, we have the following results on N_k according to [14, 25].

Lemma 3.3 *If each tag replies in a random slot among the L_k slots with probability r_k, then it holds that $N_k \sim \mathcal{N}[\mu, \sigma^2]$ for large L_k and z_k, where $\mu = L_k(1 - \frac{r_k}{L_k})^{z_k}$ and $\sigma^2 = L_k(L_k - 1)(1 - \frac{2r_k}{L_k})^{z_k} + L_k(1 - \frac{r_k}{L_k})^{z_k} - L_k{}^2(1 - \frac{r_k}{L_k})^{2z_k}$.*

Lemma 3.4 *For any $\epsilon^* > 0$, there exists some $M > 0$, such that if $z_k \geq M$ or $L_k = \hat{z}_{k|k-1} \geq M$, then it holds that*

$$\left| \mu - L_k e^{-r_k \rho} \right| \leq \epsilon^*, \tag{3.21}$$

$$\left| \sigma^2 - L_k (e^{-r_k \rho} - (1 + r_k^2 \rho)e^{-2r_k \rho}) \right| \leq \epsilon^*, \tag{3.22}$$

where $\rho = \frac{z_k}{L_k}$ is referred to as the reader load factor.

Lemmas 3.3 and 3.4 imply that in large-scale RFID systems, we can use $L_k e^{-r_k \rho}$ and $L_k(e^{-r_k \rho} - (1 + r_k^2 \rho)e^{-2r_k \rho})$ to approximate μ and σ^2.

At the end of each frame k, the reader gets a measure y_k of the idle slot frequency defined as

$$y_k = \frac{N_k}{L_k}. \tag{3.23}$$

Recall Lemma 3.3, it holds that y_k is a Normal distributed random variable specified as follows: $E[y_k] = e^{-r_k \rho}$ and $Var[y_k] = \frac{1}{L_k}(e^{-r_k \rho} - (1 + r_k^2 \rho)e^{-2r_k \rho})$. Since there are z_k tags reply in frame k with probability r_k, the probability that a slot is idle, denoted as $p(z_k)$, can be calculated as

$$p(z_k) = (1 - \frac{r_k}{L_k})^{z_k} \approx e^{-\frac{r_k z_k}{L_k}}. \tag{3.24}$$

Notice that for large z_k, $p(z_k)$ can be regarded as a continuously differentiable function of z_k.

Using the language in the Kalman filter, we can write y_k as follows:

$$y_k = p(z_k) + u_k, \tag{3.25}$$

where, based on the statistic characteristics of y_k, u_k is a Gaussian random variable with zero mean and variance

$$Var[u_k] = \frac{1}{L_k}(e^{-r_k \rho} - (1 + r_k^2 \rho)e^{-2r_k \rho}). \tag{3.26}$$

We note that u_k measures the uncertainty of y_k.

To summarise, the discrete-time model for static RFID systems is characterized by (3.20) and (3.25).

3.5.2 Tag Population Estimation Algorithm

Noticing that the system state characterised by (3.20) and (3.25) is a discrete-time nonlinear system, we thus leverage the two-step EKF described in Definition 3.1 to estimate the system state. In (3.7), the Kalman gain K_k increases with Q_k while decreases with R_k. As a result, Q_k and R_k can be used to tune the EKF such that increasing Q_k and/or decreasing R_k accelerates the convergence rate but leads to larger estimation error. In our design, we set Q_k to a constant $q > 0$ and introduce a parameter ϕ_k as follows to replace R_k to facilitate our demonstration:

$$R_k = \phi_k P_{k|k-1} C_k{}^2. \tag{3.27}$$

It can be noted from (3.7) and (3.27) that K_k is monotonously decreasing in ϕ_k, i.e., a small ϕ_k leads to quick convergence with the price of relatively high estimation error. Hence, choosing the appropriate value for ϕ_k consists of striking a balance between the convergence rate and the estimation error. In our work, we take a dynamic approach by setting ϕ_k to a small value $\underline{\phi}$ but satisfying (3.64) at the first few rounds (J rounds) of estimation to allow the system to act quickly since the estimation in the beginning phase can be very coarse. After that we set ϕ_k to a relatively high value $\overline{\phi}$ to achieve high estimation accuracy.

Now, we present our tag population estimation algorithm in Algorithm 1 where $P_{0|0}, q$ can be set to some constants straightforward since the performance mostly depends on ϕ_k and k_{max} is the pre-configured time horizon during which the system needs to be monitored. The major procedures can be summarised as:

1. *In the beginning of frame k: prediction (line 3).* The reader first predicts the number of tags based on the estimation at the end of frame $k - 1$. The predicted value is defined as $\hat{z}_{k|k-1}$. Then the reader sets the persistence probability r_k following Lemma 3.8 and z_k is set to $\hat{z}_{k|k-1}$.
2. *Line 4–5.* The reader launches the *Listen-before-talk* protocol as introduced in 3.4.1 in order to receive the feedbacks from tags.

Algorithm 1 Tag population estimation (static cases): executed by the reader

---

Require: $\underline{z}_0, P_{0|0}, q, J, L, \underline{\phi}, \overline{\phi}$, maximum number of rounds k_{max}
Ensure: Estimated tag population set $S_z = \{\hat{z}_{k|k} : k \in [0, k_{max}]\}$
1: **Initialisation:** $\hat{z}_{0|0} \leftarrow \underline{z}_0, Q_0 \leftarrow q, S_z = \{\hat{z}_{0|0}\}$
2: **for** $k = 1$ to k_{max} **do**
3:     $\hat{z}_{k|k-1} \leftarrow \hat{z}_{k-1|k-1}, L_k \leftarrow L, r_k \leftarrow 1.59 L_k / \hat{z}_{k|k-1}, P_{k|k-1} \leftarrow P_{k-1|k-1} + Q_{k-1}$
4:     Generate a new random seed Rs_k and broadcast (L_k, r_k, Rs_k)
5:     Run *Listen-before-Talk* protocol
6:     Obtain the number of idle slots N_k, and compute y_k and v_k using (3.23) and (3.8)
7:     $Q_k \leftarrow q$
8:     **if** $k \leq J$ **then**
9:         $\phi_k \leftarrow \underline{\phi}$
10:     **else**
11:         $\phi_k \leftarrow \overline{\phi}$
12:     **end if**
13:     Calculate R_k and K_k using (3.27) and (3.7)
14:     Update $\hat{z}_{k|k}$ and $P_{k|k}$ using (3.5) and (3.6)
15:     $S_z \leftarrow S_z \cup \{\hat{z}_{k|k}\}$
16: **end for**

---

3. *At the end of frame k: correction (line 6–14).* The reader computes N_k based on B_k and further calculates y_k and v_k from N_k. It then updates the prediction with the corrected estimate $\hat{z}_{k|k}$ following (3.5).

We will theoretically establish the stability and accuracy of the algorithm in Sect. 3.7.

3.6   Tag Population Estimation: Dynamic Systems

In this section, we further tackle the dynamic case where the tag population may vary during the estimation process. The objective for the dynamic systems is to promptly detect the global tag papulation change and accurately estimate the quantity of this change. To that end, we first establish the system model and then present our estimation algorithm.

3.6.1   System Dynamics and Measurement Model

In dynamic RFID systems, we can formulate the system dynamics as

$$z_{k+1} = z_k + w_k, \tag{3.28}$$

where the tag population z_{k+1} in frame $k + 1$ consists of two parts: (1) the tag population in frame k and (2) a random variable w_k which accounts for the stochastic variation of tag population resulting from the tag arrival/departure during frame k. Notice that w_k is referred to as process noise in Kalman filters and the appropriate characterisation of w_k is crucial in the design of stable Kalman filters, which will be investigated in detail later. Besides, the measurement model is the same as the static case. Hence, the discrete-time model for dynamic RFID systems can be characterized by (3.28) and (3.25).

3.6.2   Tag Population Estimation Algorithm

In the dynamic case, we leverage the two-step EKF to estimate the system state combined with the CUSUM test to further trace the tag population fluctuation.

Our main estimation algorithm is illustrated in Algorithm 2. The difference compared to the static scenario is that tag population variation needs to be detected by the CUSUM test presented in Algorithm 3 in the next subsection and the output of Algorithm 3 acts as a feedback to ϕ_k, meaning ϕ_k is no more a constant after the first J rounds as the static case due to the tag population variation. Specifically, if a

---

Algorithm 2 Tag population estimation (unified framework): executed by the reader

---

Require: $\underline{z}_0$, $P_{0|0}$, q, J, L, $\underline{\phi}$, $\overline{\phi}$, maximum number of rounds k_{max}
Ensure: Estimation set $S_z = \{\hat{z}_{k|k} : k \in [0, k_{max}]\}$
 1: **Initialisation:** $\hat{z}_{0|0} \leftarrow \underline{z}_0$, $Q_0 \leftarrow q$, $S_z = \{\hat{z}_{0|0}\}$
 2: **for** $k = 1$ to k_{max} **do**
 3:    $\hat{z}_{k|k-1} \leftarrow \hat{z}_{k-1|k-1}$, $L_k \leftarrow L$, $r_k \leftarrow 1.59L_k/\hat{z}_{k|k-1}$, $P_{k|k-1} \leftarrow P_{k-1|k-1} + Q_{k-1}$
 4:    Generate a new seed Rs_k and broadcast (L_k, Rs_k) and run *Listen-before-Talk* protocol
 5:    Obtain the number of idle slots N_k, and compute y_k and v_k using (3.23) and (3.8)
 6:    $Q_k \leftarrow q$
 7:    **if** $k \leq J$ **then**
 8:        $\phi_k \leftarrow \underline{\phi}$
 9:    **else**
10:        Execute Algorithm 3
11:        $\phi_k \leftarrow$ *output of Algorithm 3*
12:    **end if**
13:    Calculate R_k and K_k using (3.27) and (3.7), and update $\hat{z}_{k|k}$ and $P_{k|k}$ using (3.5) and (3.6)
14:    $S_z \leftarrow S_z \cup \{\hat{z}_{k|k}\}$
15: **end for**

Algorithm 3 CUSUM test: executed by the reader in frame k

Require: Υ, θ
Ensure: ϕ_k
 1: **Initialisation:** $g_0^+ \leftarrow 0$, $g_0^- \leftarrow 0$
 2: Compute Φ_k using equation (3.29)
 3: $g_k^+ \leftarrow$ (3.30), $g_k^- \leftarrow$ (3.31)
 4: **if** $g_k^+ > \theta$ or $g_k^- < -\theta$ **then**
 5:    $\delta \leftarrow 1$, $\phi_k \leftarrow \varphi_1(\delta)$, $g_k^+ \leftarrow 0$, $g_k^- \leftarrow 0$
 6: **else**
 7:    $\delta \leftarrow 0$, $\phi_k \leftarrow \varphi_1(\delta)$
 8: **end if**
 9: Return ϕ_k

---

change on the tag population is detected in frame k, ϕ_k is set to $\underline{\phi}$ to quickly react to the change, otherwise ϕ_k sticks to $\overline{\phi}$ to stabilize the estimation. The overall structure of the estimation algorithm is illustrated in Fig. 3.1. We note that in the case where z_k is constant, Algorithm 2 degenerates to Algorithm 1.

3.6.3 Detecting Tag Population Change: CUSUM Test

The CUSUM Detection Framework We leverage the CUSUM test to detect the change of tag population and further adjust ϕ_k. CUSUM test is a sequential analysis technique typically used for change detection [26]. It is shown to be asymptotically optimal in the sense of the minimum detection time subject to a fixed worst-case expected false alarm rate [27].

Fig. 3.1 Estimation
algorithm diagram: dashed
box indicates the EKF

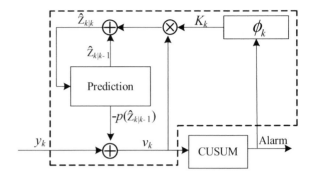

In the context of dynamic tag population detection, the reader monitors the innovation process $v_k = y_k - p(\hat{z}_{k|k-1})$. If the number of the tags population is constant, v_k equals to u_k which is a Gaussian process with zero mean. In contrast, upon the system state changes, i.e., tag population changes, v_k drifts away from the zero mean. In our design, we use Φ_k as a normalised input to the CUSUM test by normalising v_k with its estimated standard variance, specified as follows:

$$\Phi_k = \frac{v_k}{\sqrt{(P_{k|k-1} + Q_{k-1})C_k^2 + Var[u_k]\big|_{z_k=\hat{z}_{k|k-1}}}}. \tag{3.29}$$

The reader further updates the CUSUM statistics g_k^+ and g_k^- as follows:

$$g_k^+ = \max\{0, g_{k-1}^+ + \Phi_k - \Upsilon\}, \tag{3.30}$$

$$g_k^- = \min\{0, g_{k-1}^- + \Phi_k + \Upsilon\}, \tag{3.31}$$

$$g_k^+ = g_k^- = 0, \text{ if } \delta = 1, \tag{3.32}$$

where $g_0^+ = 0$ and $g_0^- = 0$. And $\Upsilon \geq 0$, referred to as reference value, is a filter design parameter indicating the sensitivity of the CUSUM test to the fluctuation of Φ_k, Moreover, by δ we define an indicator flag indicating tag population change:

$$\delta = \begin{cases} 1 & \text{if } g_k^+ > \theta \text{ or } g_k^- < -\theta, \\ 0 & \text{otherwise}, \end{cases} \tag{3.33}$$

where $\theta > 0$ is a pre-specified CUSUM threshold.

The detailed procedure of the change detection is illustrated in Algorithm 3 where $\varphi_1(\delta)$ is used to assign the value to ϕ_k according to whether the system state changes and is shown in (3.37).

Parameter Tuning in CUSUM Test The choice of the threshold θ and the drift parameter Υ has a directly impact on the performance of the CUSUM test in terms of detection delay and false alarm rate. Formally, the average running length (ARL) $L(\mu^*)$ is used to denote the duration between two actions [28]. For a large θ, $L(\mu^*)$ can be approximated as [2]

$$L(\mu^*) = \begin{cases} \Theta(\theta), & \text{if } \mu^* \neq 0, \\ \Theta(\theta^2), & \text{if } \mu^* = 0, \end{cases} \tag{3.34}$$

where μ^* denotes the mean of the process Φ_k.

In our context, ARL corresponds to the mean time between two false alarms in the static case and the mean detection delay of the tag population change in the dynamic case. It is easy to see from (3.34) that a higher value of θ leads to lower false alarm rate at the price of longer detection delay. Therefore, the choices of θ and Υ consists of a tradeoff between the false alarm rate and the detection delay.

Recall that Φ_k can be approximated to a white noise process, i.e, $\Phi_k \sim \mathcal{N}[\mu^*, \sigma^{*2}]$ with $\mu^* = 0$, $\sigma^* = 1$ if the system state does not change. Generically, as recommended in [29], setting θ and Υ as follows achieves good ARL from the engineering perspective.

$$\theta = 4\sigma^*, \tag{3.35}$$

$$\Upsilon = \mu^* + 0.5\sigma^*. \tag{3.36}$$

In the CUSUM framework, we set ϕ_k by $\varphi_1(\delta)$ as follows:

$$\varphi_1(\delta) = \begin{cases} \underline{\phi}, & \text{if } \delta = 1, \\ \overline{\phi}, & \text{if } \delta = 0. \end{cases} \tag{3.37}$$

The rationale is that once a change on the tag population is detected in frame k, ϕ_k is set to $\underline{\phi}$ to quickly react to the change, while ϕ_k sticks to $\overline{\phi}$ when no system change is detected.

3.7 Performance Analysis

In this section, we establish the stability and the accuracy of our estimation algorithms for both static and dynamic cases.

---

[2]For two variables X, Y, asymptotic notation $X = \Theta(Y)$ implies that there exist positives c_1, c_2 and x_0 such that for $\forall X > x_0$, it follows that $c_1 X \leq Y \leq c_2 X$.

3.7.1 Static Case

Our analysis is composed of two steps. We first derive the estimation error and then establish the stability and the accuracy of Algorithm 1 in terms of the boundedness of estimation error.

Computing Estimation Error We first approximate the non-linear discrete system by a linear one. To that end, as the function $p(z_k)$ is continuously differentiable at $z_k = \hat{z}_{k|k-1}$, using the Taylor expansion, we have

$$p(z_k) = p(\hat{z}_{k|k-1}) + C_k(z_k - \hat{z}_{k|k-1}) + \chi(z_k, \hat{z}_{k|k-1}), \tag{3.38}$$

where

$$C_k = -\frac{r_k \rho}{e^{r_k \rho} \hat{z}_{k|k-1}}, \tag{3.39}$$

$$\chi(z_k, \hat{z}_{k|k-1}) = \sum_{j=2}^{\infty} \frac{1}{e^{r_k \rho} j!} (r_k \rho - \frac{r_k \rho z_k}{\hat{z}_{k|k-1}})^j. \tag{3.40}$$

Regarding the convergence of $\chi(z_k, \hat{z}_{k|k-1})$ in (3.40), assume that

$$z_k = a'_k \hat{z}_{k|k-1}, \tag{3.41}$$

we can obtain the following boundedness of the residual for the case $|a'_k - 1| < \frac{1}{r_k \rho}$:

$$|\chi(z_k, \hat{z}_{k|k-1})| = \frac{(r_k\rho)^2(\hat{z}_{k|k-1} - z_k)^2}{e^{r_k\rho}\hat{z}_{k|k-1}^2} \sum_{j=0}^{\infty} \frac{(r_k\rho)^j}{(j+2)!} \left|1 - \frac{z_k}{\hat{z}_{k|k-1}}\right|^j \tag{3.42}$$

$$\leq \frac{(r_k\rho)^2(\hat{z}_{k|k-1} - z_k)^2}{2e^{(r_k\rho)}\hat{z}_{k|k-1}^2[1 - |(r_k\rho)(1 - \frac{z_k}{\hat{z}_{k|k-1}})|]} \leq \frac{(r_k\rho)^2(\hat{z}_{k|k-1} - z_k)^2}{2e^{(r_k\rho)}a_k\hat{z}_{k|k-1}^2}, \tag{3.43}$$

where

$$a_k = 1 - (r_k\rho)|1 - a'_k|. \tag{3.44}$$

Recall the definition of the estimation error in (3.10) and using (3.20), (3.3) and (3.5), we can derive the estimation error $e_{k+1|k}$ as follows:

$$\begin{aligned} e_{k+1|k} &= z_{k+1} - \hat{z}_{k+1|k} = z_k - \hat{z}_{k|k} = z_k - \hat{z}_{k|k-1} \\ &\quad - K_k \left[C_k(z_k - \hat{z}_{k|k-1}) + \chi(z_k, \hat{z}_{k|k-1}) + u_k \right] \\ &= (1 - K_k C_k)e_{k|k-1} + s_k + m_k, \end{aligned} \tag{3.45}$$

where s_k and m_k are defined as

$$s_k = -K_k u_k, \tag{3.46}$$

$$m_k = -K_k \chi (z_k, \hat{z}_{k|k-1}). \tag{3.47}$$

Boundedness of Estimation Error Having derived the dynamics of the estimation error, we now state the main result on the stochastic stability and accuracy of Algorithm 1.

Theorem 3.1 *Consider the discrete-time stochastic system given by (3.20) and (3.25) and Algorithm 1, the estimation error $e_{k|k-1}$ defined by (3.10) is exponentially bounded in mean square and bounded w.p.o., if the following conditions hold:*

1. *there are positive numbers $\underline{q}, \overline{q}, \underline{\phi}$ and $\overline{\phi}$ such that the bounds on Q_k and ϕ_k are satisfied for every $k \geq 0$, as in*

$$\underline{q} \leq Q_k \leq \overline{q}, \tag{3.48}$$

$$\underline{\phi} \leq \phi_k \leq \overline{\phi}, \tag{3.49}$$

2. *The initialization must follow the rules*

$$P_{0|0} > 0, \tag{3.50}$$

$$|e_{1|0}| \leq \epsilon \tag{3.51}$$

with positive real number $\epsilon > 0$.

Remark By referring to the design objective posed in Sect. 3.4, Theorem 3.1 prove the following properties of our estimation algorithm:

- the estimation error of our algorithm is bounded in mean square and the relative estimation error tends to zero;
- the estimated population size converges to the real value with exponential rate.

Moreover, the conditions in Theorem 3.1 can be interpreted as follows:

1. The inequalities (3.48) and (3.49) can be satisfied by the configuring the correspondent parameters in Algorithm 1, which guarantees the boundedness of the pseudo-covariance $P_{k|k-1}$ as shown later.
2. The inequality (3.50) consists of establishing positive $P_{k|k-1}$ for every $k \geq 1$.
3. As a sufficient condition for stability, the upper bound ϵ may be too stringent. As shown in the simulation results, stability is still ensured even with a relatively large ϵ.

Before the proof of Theorem 3.1, we first state several auxiliary lemmas to streamline the proof and show how to apply these lemmas to prove Theorem 3.1 subsequently.

Lemma 3.5 *Under the conditions of Theorem 3.1, if $P_{0|0} > 0$, there exist $\underline{p}_k, \overline{p}_k > 0$ such that the pseudo-covariance $P_{k|k-1}$ is bounded for every $k \geq 1$, i.e.,*

$$\underline{p}_k \leq P_{k|k-1} \leq \overline{p}_k. \tag{3.52}$$

Proof Recall (3.4) and (3.6), we have $P_{k|k-1} \geq Q_{k-1}$, and

$$P_{k|k-1} = P_{k-1|k-2}(1 - K_{k-1}C_{k-1}) + Q_{k-1}$$

$$= P_{k-1|k-2}\left(1 - \frac{P_{k-1|k-2}C_{k-1}^2}{P_{k-1|k-2}C_{k-1}^2 + R_{k-1}}\right) + Q_{k-1}. \tag{3.53}$$

Following the design of R_k in (3.27) and by iteration, we further get

$$P_{k|k-1} = P_{k-1|k-2}\left(1 - \frac{1}{1 + \phi_{k-1}}\right) + Q_{k-1}$$

$$= P_{1|0}\prod_{i=1}^{k-1}\left(1 - \frac{1}{1 + \phi_i}\right) + \sum_{i=0}^{k-2}Q_i\prod_{j=i}^{k-2}\left(1 - \frac{1}{1 + \phi_{j+1}}\right) + Q_{k-1}.$$

Since ϕ_k and Q_k are controllable parameters, we can set $\phi_k \leq \overline{\phi}$ and $Q_k \leq \overline{q}$ for every $k \geq 0$ in Algorithm 1, where $\overline{\phi}, \overline{q} > 0$. Consequently, we have

$$P_{k|k-1} \leq P_{1|0}\left(1 - \frac{1}{1 + \overline{\phi}}\right)^{k-1} + \overline{q}\sum_{j=1}^{k-1}\left(1 - \frac{1}{1 + \overline{\phi}}\right)^j + Q_{k-1}$$

$$\leq (P_{0|0} + Q_0)\left(1 - \frac{1}{1 + \overline{\phi}}\right)^{k-1} + \overline{q}\overline{\phi} + Q_{k-1} \tag{3.54}$$

Let $\overline{p}_k = ((P_{0|0} + Q_0)\left(1 - \frac{1}{1+\overline{\phi}}\right)^{k-1} + \overline{q}\overline{\phi} + Q_{k-1}$ and $\underline{p}_k = Q_{k-1}$, we have $\underline{p}_k \leq P_{k|k-1} \leq \overline{p}_k$. □

Lemma 3.6 *Let $\alpha_k \triangleq \frac{1}{1+\phi_k}$, it holds that*

$$\frac{(1 - K_kC_k)^2}{P_{k+1|k}}e_{k|k-1}^2 \leq (1 - \alpha_k)\frac{e_{k|k-1}^2}{P_{k|k-1}}, \quad \forall k \geq 1. \tag{3.55}$$

Proof From (3.53), we have

$$P_{k+1|k} = P_{k|k-1}(1 - K_kC_k) + Q_k \geq P_{k|k-1}(1 - K_kC_k). \tag{3.56}$$

By substituting it into the left-hand side of (3.55) and using the fact that $R_k = \phi_k P_{k|k-1} C_k{}^2$ for every $k \geq 1$, we get

$$\frac{(1 - K_k C_k)^2}{P_{k+1|k}} e_{k|k-1}^2 \leq \frac{(1 - K_k C_k)^2}{P_{k|k-1}(1 - K_k C_k)} e_{k|k-1}^2 \leq (1 - K_k C_k) \frac{e_{k|k-1}^2}{P_{k|k-1}}$$

$$\leq \left(1 - \frac{1}{1 + \phi_k}\right) \frac{e_{k|k-1}^2}{P_{k|k-1}}.$$

We are thus able to prove (3.55).   □

Lemma 3.7 *Let* $b_k \triangleq \dfrac{r_k \rho (4 a_k \phi_k + 1 - a_k)}{4 a_k^2 \phi_k (1 + \phi_k) \hat{z}_{k|k-1} P_{k|k-1}}$, *it holds that*

$$\frac{m_k[2(1 - K_k C_k) e_{k|k-1} + m_k]}{P_{k+1|k}} \leq b_k |\hat{z}_{k|k-1} - z_k|^3. \tag{3.57}$$

Proof From (3.47), we get the following expansion

$$\frac{m_k[2(1 - K_k C_k) e_{k|k-1} + m_k]}{P_{k+1|k}} = \frac{1}{P_{k+1|k}} \frac{-P_{k|k-1} C_k}{P_{k|k-1} C_k{}^2 + R_k} \chi(z_k, \hat{z}_{k|k-1})$$

$$\cdot \left[2 \left(1 - \frac{P_{k|k-1} C_k{}^2}{P_{k|k-1} C_k{}^2 + R_k}\right) e_{k|k-1} - \frac{P_{k|k-1} C_k}{P_{k|k-1} C_k{}^2 + R_k} \chi(z_k, \hat{z}_{k|k-1})\right]. \tag{3.58}$$

It then follows from (3.39), (3.41) and (3.56) that

$$\frac{m_k[2(1 - K_k C_k) e_{k|k-1} + m_k]}{P_{k+1|k}}$$

$$\leq \frac{1}{P_{k|k-1}(1 - K_k C_k)} \frac{-P_{k|k-1} C_k}{P_{k|k-1} C_k{}^2 + R_k} \frac{(r_k \rho)^2 (\hat{z}_{k|k-1} - z_k)^2}{2 e^{r_k \rho} a_k \hat{z}_{k|k-1}^2}$$

$$\cdot \left[2 \left(1 - \frac{P_{k|k-1} C_k{}^2}{P_{k|k-1} C_k{}^2 + R_k}\right) |\hat{z}_{k|k-1} - z_k|\right.$$

$$\left. - \frac{P_{k|k-1} C_k}{P_{k|k-1} C_k{}^2 + R_k} \frac{(r_k \rho)^2 (\hat{z}_{k|k-1} - z_k)^2}{2 e^{1.59} a_k \hat{z}_{k|k-1}^2}\right]$$

$$\leq \frac{r_k \rho (4 a_k \phi_k + 1 - a_k)}{4 a_k^2 \phi_k (1 + \phi_k) \hat{z}_{k|k-1} P_{k|k-1}} |\hat{z}_{k|k-1} - z_k|^3.$$

We are thus able to prove (3.57).

                                                                                                                        □

Lemma 3.8 $E\left[\frac{s_k^2}{P_{k+1|k}}\Big|e_{k|k-1}\right] \leq \frac{2.46\hat{z}_{k|k-1}}{\phi_k(1+\phi_k)r_k P_{k|k-1}}$ *when* $r_k\rho = 1.59$.

Proof From (3.46), we have $E\left[\frac{s_k^2}{P_{k+1|k}}\Big|e_{k|k-1}\right] = \frac{K_k^2 E[u_k^2]}{P_{k+1|k}}$. With (3.7), (3.26) and (3.56), we have

$$E\left[\frac{s_k^2}{P_{k+1|k}}\Big|e_{k|k-1}\right] \leq \frac{e^{2r_k\rho}\hat{z}_{k|k-1}}{\phi_k(1+\phi_k)P_{k|k-1}\rho r_k^2}(e^{-r_k\rho} - (1+r_k^2\rho)e^{-2r_k\rho}).$$

Since item $E\left[\frac{s_k^2}{P_{k+1|k}}\Big|e_{k|k-1}\right]$ influences the estimation accuracy, we set the optimal persistence probability to minimize this item. Denote $\Lambda(r_k) = \frac{e^{2r_k\rho}}{r_k^2}(e^{-r_k\rho} - (1+r_k^2\rho)e^{-2r_k\rho})$, we have

$$\frac{d\Lambda}{dr_k} = \frac{(r_k\rho - 2)e^{r_k\rho} + 2}{r_k^3}.$$

Since $r_k\rho > 0$ and $\frac{d((r_k\rho-2)e^{r_k\rho}+2)}{dr_k\rho} = (r_k\rho - 1)e^{r_k\rho}$ which is greater zero if $r_k\rho > 1$ and is smaller than zero if $r_k\rho < 1$, and 1) if $r_k\rho = 1$, $\frac{d\Lambda}{dr_k} < 0$; 2) if $r_k\rho = 0$, $\frac{d\Lambda}{dr_k} = 0$; 3) if $r_k\rho = 2$, $\frac{d\Lambda}{dr_k} > 0$, there exists a unique solution $r_k\rho \in (1,2)$ for $\frac{d\Lambda}{dr_k} = 0$ such that $\Lambda(r_k)$ is minimized. Searching in $(1,2)$, we find the optimal $r_k\rho = 1.59$. Therefore, we can obtain that

$$E\left[\frac{s_k^2}{P_{k+1|k}}\Big|e_{k|k-1}\right] \leq \frac{2.46\hat{z}_{k|k-1}}{\phi_k(1+\phi_k)r_k P_{k|k-1}} \triangleq \xi_k, \tag{3.59}$$

which completes the proof.   □

Armed with the above auxiliary lemmas, we next prove Theorem 3.1.

Proof of Theorem 3.1 First, we construct the following Lyapunov function to define the stochastic process:

$$V_k(e_{k|k-1}) = \frac{e_{k|k-1}^2}{P_{k|k-1}},$$

which satisfies (3.4) and (3.50) as $P_{k|k-1} > 0$.

Next, we use Lemma 3.2 to develop the proof. Because it follows from Lemma 3.5 that the properties (3.16) and (3.17) in Lemma 3.2 are satisfied, the main task left is to prove (3.18).

From (3.45), expanding $V_{k+1}(e_{k+1|k})$ leads to

$$
\begin{aligned}
V_{k+1}(e_{k+1|k}) &= \frac{e_{k+1|k}^2}{P_{k+1|k}} = \frac{[(1 - K_k C_k)e_{k|k-1} + s_k + m_k]^2}{P_{k+1|k}} = \frac{(1 - K_k C_k)^2}{P_{k+1|k}} e_{k|k-1}^2 \\
&\quad + \frac{m_k[2(1 - K_k C_k)e_{k|k-1} + m_k]}{P_{k+1|k}} + \frac{2s_k[(1 - K_k C_k)e_{k|k-1} + m_k]}{P_{k+1|k}} \\
&\quad + \frac{s_k^2}{P_{k+1|k}}.
\end{aligned}
$$

Furthermore, by Lemmas 3.6, 3.7 and 3.8 and some algebraic operations, we have

$$
E\left[V_{k+1}(e_{k+1|k})|e_{k|k-1}\right] - V_k(e_{k|k-1}) \le -\alpha_k V_k(e_{k|k-1}) + b_k |e_{k|k-1}|^3 + \xi_k. \tag{3.60}
$$

To obtain the same formation with (3.18), we further proceed to bound the second term in b_k in (3.60) as follows:

$$
b_k |e_{k|k-1}|^3 \le \varsigma \alpha_k V_k(e_{k|k-1}), \tag{3.61}
$$

where $0 < \varsigma < 1$ is preset controllable parameter. To prove the above inequality, we need to prove $|e_{k|k-1}| \le \frac{4\varsigma a_k^2 \phi_k \hat{z}_{k|k-1}}{1.59(4a_k \phi_k + 1.59|a_k' - 1|)}$. Since $|e_{k|k-1}| = |a_k' - 1|\hat{z}_{k|k-1}$, it suffices to show

$$
|a_k' - 1| \le \frac{4\varsigma a_k^2 \phi_k}{1.59(4a_k \phi_k + 1.59|a_k' - 1|)}, \tag{3.62}
$$

which is equivalent to $(1 - 4\phi_k - 4\phi_k \varsigma)a_k^2 + (4\phi_k - 2)a_k + 1 \le 0$ because of (3.44). With some algebraic operations, we obtain (1) $\frac{1 - 2\phi_k - 2\sqrt{\phi_k(\phi_k + \varsigma)}}{1 - 4\phi_k(1 + \varsigma)} < a_k \le 1$, if $\phi_k < \frac{1}{4(1+\varsigma)}$; and (2) $\frac{2\phi_k - 1 + 2\sqrt{\phi_k(\phi_k + \varsigma)}}{4\phi_k(1+\varsigma) - 1} \le a_k \le 1$, if $\phi_k > \frac{1}{4(1+\varsigma)}$; and (3) $\frac{1+\varsigma}{1+2\varsigma} \le a_k \le 1$, if $\phi_k = \frac{1}{4(1+\varsigma)}$. Since it holds that $\frac{2\phi_k - 1 + 2\sqrt{\phi_k(\phi_k + \varsigma)}}{4\phi_k(1+\varsigma) - 1} < \frac{1+\varsigma}{1+2\varsigma}$ for every ς and $\frac{2\phi_k - 1 + 2\sqrt{\phi_k(\phi_k + \varsigma)}}{4\phi_k(1+\varsigma) - 1}$ will decrease monotonically to $\frac{1}{1+\varsigma}$ for a large ϕ_k, we have in the worst case for $\phi_k \ge \frac{1}{4(1+\varsigma)}$,

$$
\frac{1+\varsigma}{1+2\varsigma} \le a_k \le 1. \tag{3.63}
$$

It follows from the analysis that if we set

$$
\phi_k \ge \frac{1}{4(1+\varsigma)}, \tag{3.64}
$$

(3.62) can be satisfied. Moreover, it holds that

$$|a'_k - 1| \leq \frac{0.63\varsigma}{1 + 2\varsigma}. \tag{3.65}$$

That is,

$$|e_{k|k-1}| \leq \epsilon_k, \tag{3.66}$$

where $\epsilon_k \triangleq \frac{0.63\varsigma}{1+2\varsigma}\hat{z}_{k|k-1}$. By setting ϕ_k in (3.64), for $|e_{k|k-1}| \leq \epsilon_k$, we thus have

$$E\left[V_{k+1}(e_{k+1|k})|e_{k|k-1}\right] - V_k(e_{k|k-1}) \leq -(1-\varsigma)\alpha_k V_k(e_{k|k-1}) + \xi_k. \tag{3.67}$$

Therefore, we are able to apply Lemma 3.2 to prove Theorem 3.1 by setting $\epsilon = \frac{0.63\varsigma}{1+2\varsigma}\hat{z}_{1|0}$, $\beta^* = \frac{1}{Q_0}$, $\alpha_k^* = (1-\varsigma)\alpha_k$, $\beta_k = \frac{1}{p_k}$ and $\tau_k = \xi_k$. □

Remark Theorem 3.1 also holds in the sense of Lemma 3.1 (the off-line version of Lemma 3.2) by setting the parameters in (3.15) as $\overline{\beta} = \frac{1}{Q_0}$, $\alpha = \frac{1-\varsigma}{1+\overline{\phi}} \leq \alpha_k^*$, $\underline{\beta} = (P_{0|0} + Q_0 + \overline{q}(\overline{\phi} + 1)) \geq \overline{p}_k$, and $\tau = \frac{Q_0\hat{z}_{max}}{\overline{\phi}(1+\overline{\phi})} \geq \xi_k$, where $\hat{z}_{max}$ is the maximum estimate.

We conclude the analysis on the performance of our estimation algorithm for the static case with a more profound investigation on the evolution of the estimation error $|e_{k|k-1}|$. More specifically, we can distinguish three regions:

- *Region 1:* $\sqrt{\frac{2.46M\hat{z}_{k|k-1}}{\phi_k(M-1)r_k(1-\varsigma)}} \leq |e_{k|k-1}| \leq \epsilon_k$. By substituting the condition into the right hand side of (3.67), we obtain: $-(1-\varsigma)\alpha_k V_k(e_{k|k-1}) + \xi_k \leq -\frac{(1-\varsigma)\alpha_k}{M}V_k(e_{k|k-1})$, where $M > 1$ is a positive constant and can be set beforehand. It then follows that

$$E\left[V_{k+1}(e_{k+1|k})|e_{k|k-1}\right] \leq \left(1 - \frac{(1-\varsigma)\alpha_k}{M}\right)V_k(e_{k|k-1}).$$

Consequently, we can bound $E[e_{k|k-1}^2]$ as:

$$E[e_{k|k-1}^2] \leq \frac{\overline{p}_k}{Q_0}E[e_{1|0}^2]\prod_{i=1}^{k-1}(1 - \alpha_i^*) \tag{3.68}$$

with $\alpha_k^* = \frac{(1-\varsigma)\alpha_k}{M}$. It can then be noted that $E[e_{k|k-1}^2] \to 0$ at an exponential rate as $k \to \infty$.
- *Region 2:* $\sqrt{\frac{2.46\hat{z}_{k|k-1}}{\phi_k r_k(1-\varsigma)}} \leq |e_{k|k-1}| < \sqrt{\frac{2.46M\hat{z}_{k|k-1}}{\phi_k(M-1)r_k(1-\varsigma)}}$. In this case, we have

$$-\frac{(1-\varsigma)\alpha_k}{M}V_k(e_{k|k-1}) < -(1-\varsigma)\alpha_k V_k(e_{k|k-1}) + \xi_k \leq 0.$$

It then follows from Lemma 3.2 that

$$E[e_{k|k-1}^2] \leq \frac{\overline{P}_k}{Q_0} E[e_{1|0}^2] \prod_{i=1}^{k-1} (1 - \alpha_i^*) + \overline{P}_k \sum_{i=1}^{k-2} \xi_{k-i-1} \prod_{j=1}^{i} (1 - \alpha_{k-j}^*).$$

(3.69)

Hence, when $k \to \infty$, $E[e_{k|k-1}^2]$ converges at exponential rate to $\overline{P}_k \sum_{i=1}^{k-2} \xi_{k-i-1} \cdot \prod_{j=1}^{i} (1 - \alpha_{k-j}^*) \sim \Theta(\hat{z}_{k|k-1})$, which is decoupled with the initial estimation error and it thus holds $\frac{E[e_{k|k-1}]}{z_k} = \Theta\left(\frac{1}{\sqrt{z_k}}\right) \to 0$ when $z_k \to \infty$.

- *Region 3*: $0 \leq |e_{k|k-1}| < \sqrt{\frac{2.46\hat{z}_{k|k-1}}{\phi_k r_k (1-\varsigma)}}$. In this case, we can show that the right hand side of (3.67) is positive, i.e., $-(1-\varsigma)\alpha_k V_k(e_{k|k-1}) + \xi_k > 0$. It also follows from Lemma 3.2 that

$$E[e_{k|k-1}^2] \leq \frac{\overline{P}_k}{Q_0} E[e_{1|0}^2] \prod_{i=1}^{k-1} (1 - \alpha_i^*) + \overline{P}_k \sum_{i=1}^{k-2} \xi_{k-i-1} \prod_{j=1}^{i} (1 - \alpha_{k-j}^*).$$

(3.70)

Hence, when $k \to \infty$, $E[e_{k|k-1}^2]$ converges exponentially to $\overline{P}_k \sum_{i=1}^{k-2} \xi_{k-i-1} \cdot \prod_{j=1}^{i} (1 - \alpha_{k-j}^*) \sim \Theta(\hat{z}_{k|k-1})$, which is decoupled with the initial estimation error and it thus holds $\frac{E[e_{k|k-1}]}{z_k} \leq \Theta\left(\frac{1}{\sqrt{z_k}}\right) \to 0$ when $z_k \to \infty$.

Combining the above three regions, we get the following results on the convergence of the expected estimation error $E[e_{k|k-1}]$: (1) if the estimation error is small (Region 3), it will converge to a value smaller than $\Theta(\sqrt{\hat{z}_{k|k-1}})$ as analysed in Region 3; (2) if the estimation error is larger (Region 1), it will decrease as analysed in Region 1 and fall into either Region 2 or Region 3 where $E[e_{k|k-1}] \leq \Theta(\sqrt{\hat{z}_{k|k-1}})$ such that the relative estimation error $\frac{E[e_{k|k-1}]}{z_k} \to 0$ when $z_k \to \infty$.

3.7.2 Dynamic Case

Our analysis on the stability of Algorithm 2 for the dynamic case is also composed of two steps. First, we derive the estimation error. Second, we establish the stability and the accuracy of Algorithm 2 in terms of the boundedness of estimation error.

We first derive the dynamics of the estimation error as follows:

$$e_{k+1|k} = (1 - K_k C_k)e_{k|k-1} + s_k + m_k,$$

(3.71)

which differs from the static case (3.45) in s_k. In the dynamic case, we have

$$s_k = w_k - K_k u_k$$

(3.72)

Next, we show the boundedness of the estimation error in Theorem 3.2.

Theorem 3.2 *Under the conditions of Theorem 1, consider the discrete-time stochastic system given by (3.28) and (3.25) and Algorithm 2, if there exist time-varying positive real number λ_k, $\sigma_k > 0$ such that*

$$E[w_k] \leq \lambda_k, \tag{3.73}$$

$$E[w_k^2] \leq \sigma_k, \tag{3.74}$$

then the estimation error $e_{k|k-1}$ defined by (3.10) is exponentially bounded in mean square and bounded w.p.o..

Remark Note that the condition $E[w_k] \leq \lambda_k$ always holds for $E[w_k] < 0$, we thus focus on the case that $E[w_k] \geq 0$. In the proof, the explicit formulas of λ_k and σ_k are derived. As in the static case, the conditions may be too stringent such that the results still hold even if the conditions are not satisfied, as illustrated in the simulations.

The proof of Theorem 3.2 is also based on Lemmas 3.6, 3.7 and 3.8, but due to the introduction of w_k into s_k, we need another two auxiliary lemmas on $E[s_k]$ and $E[s_k^2]$.

Lemma 3.9 *If $E[w_k] \geq 0$, then there exists a time-varying real number $d_k > 0$ such that*

$$E\left[\frac{2s_k[(1 - K_k C_k)e_{k|k-1} + m_k]}{P_{k+1|k}}\Big|e_{k|k-1}\right] \leq d_k |e_{k|k-1}| E[w_k].$$

Proof When $E[w_k] \geq 0$, from $E[v_k] = 0$, (3.41), (3.56) and the independence between w_k and $e_{k|k-1}$, we can derive

$$E\left[\frac{2s_k[(1 - K_k C_k)e_{k|k-1} + m_k]}{P_{k+1|k}}\Big|e_{k|k-1}\right]$$

$$\leq 2E[w_k]\frac{1 + \phi_k}{\phi_k P_{k|k-1}}\left[\frac{\phi_k |e_{k|k-1}|}{1 + \phi_k} + \frac{1.59|e_{k|k-1}|^2}{2a_k(1 + \phi_k)\hat{z}_{k|k-1}}\right]$$

$$\leq E[w_k]\frac{2a_k \phi_k + (1 - a_k)}{a_k \phi_k P_{k|k-1}}|e_{k|k-1}|.$$

We thus complete the proof by setting

$$d_k = \frac{2a_k \phi_k + (1 - a_k)}{a_k \phi_k P_{k|k-1}}. \tag{3.75}$$

□

Lemma 3.10 *There exists a time-varying parameter $\xi_k^* > 0$ such that*
$E\left[\frac{s_k^2}{P_{k+1|k}}|e_{k|k-1}\right] \leq \xi_k^*$.

Proof By (3.72), we have $s_k^2 = w_k^2 - 2K_k w_k u_k + K_k^2 u_k^2$. Since w_k and u_k are uncorrelated and $e_{k|k-1}$ does not depend on either w_k or u_k, we have

$$E\left[\frac{s_k^2}{P_{k+1|k}}|e_{k|k-1}\right] = \frac{E[w_k^2]}{P_{k+1|k}} + \frac{K_k^2 E[u_k^2]}{P_{k+1|k}}. \tag{3.76}$$

Substituting (3.7), (3.56) and using Lemma 3.8, noticing that $E[u_k] = 0$, we get

$$E\left[\frac{s_k^2}{P_{k+1|k}}|e_{k|k-1}\right] \leq \frac{1 + \phi_k}{\phi_k P_{k|k-1}} E[w_k^2] + \frac{2.46\hat{z}_{k|k-1}}{\phi_k(1 + \phi_k)r_k P_{k|k-1}}.$$

Finally, by setting ξ_k^* as

$$\xi_k^* = \frac{1 + \phi_k}{\phi_k P_{k|k-1}} E[w_k{}^2] + \frac{2.46\hat{z}_{k|k-1}}{\phi_k(1 + \phi_k)r_k P_{k|k-1}}, \tag{3.77}$$

we complete the proof.   □

Armed with the above lemmas, we next prove Theorem 3.2 by utilizing the same method with the proof of Theorem 3.1.

Proof of Theorem 3.2 Recall (3.46) and (3.72), we notice that the only difference between the estimation errors of Algorithms 2 and 1 is s_k. Therefore, it suffices to study the impact of w_k on $V_k(e_{k|k-1})$.

It follows from Lemmas 3.6, 3.7, 3.8, 3.9 and 3.10 that

$$E\left[V_{k+1}(e_{k+1|k})|e_{k|k-1}\right] - V_k(e_{k|k-1}) \leq -\alpha_k V_k(e_{k|k-1}) + b_k|e_{k|k-1}|^3$$
$$+ d_k|e_{k|k-1}|E[w_k] + \xi_k^*.$$

Furthermore, bounding the second item in b_k as (3.61) and given ϕ_k in (3.64), yields

$$E\left[V_{k+1}(e_{k+1|k})|e_{k|k-1}\right] - V_k(e_{k|k-1}) \leq -(1 - \varsigma)\alpha_k V_k(e_{k|k-1})$$
$$+ d_k|e_{k|k-1}|E[w_k] + \xi_k^*$$

for $|e_{k|k-1}| \leq \epsilon_k$.

And we can thus prove Theorem 3.2 by setting $\epsilon = \frac{0.63\varsigma}{1+2\varsigma}\hat{z}_{1|0}$, $\beta^* = \frac{1}{Q_0}$, $\alpha_k^* = (1 - \varsigma)\alpha_k$, $\tau_k = \xi_k^* + d_k|e_{k|k-1}|\lambda_k$ and $\beta_k = \frac{1}{P_k}$.   □

We conclude the analysis on the performance of our estimation algorithm for the dynamic case with a more profound investigation on the evolution of the estimation error $|e_{k|k-1}|$ and derive the explicit formulas for λ_k and σ_k. More specifically, we can distinguish three regions:

- *Region 1*: $\sqrt{\frac{9.84M\hat{z}_{k|k-1}}{\phi_k(M-1)r_k(1-\varsigma)}} \leq |e_{k|k-1}| \leq \epsilon_k$. In this case, the objective is to achieve

$$E\left[V_{k+1}(e_{k+1|k})|e_{k|k-1}\right] - V_k(e_{k|k-1}) \leq -\frac{1}{M}(1-\varsigma)\alpha_k V_k(e_{k|k-1}) \tag{3.78}$$

so that $E[e_{k|k-1}^2]$ is bounded as

$$E[e_{k|k-1}^2] \leq \frac{\overline{p_k}}{Q_0} E[e_{1|0}^2] \prod_{i=1}^{k-1}(1-\alpha_i^*). \tag{3.79}$$

That is, it should hold that $d_k|e_{k|k-1}|E[w_k] + \xi_k^* \leq \frac{M-1}{M}(1-\varsigma)\alpha_k V_k(e_{k|k-1})$. To that end, we firstly let the following inequalities hold

$$\begin{cases} d_k|e_{k|k-1}|E[w_k] \leq \frac{M-1}{2M}(1-\varsigma)\alpha_k V_k(e_{k|k-1}), \\ \xi_k^* \leq \frac{M-1}{2M}(1-\varsigma)\alpha_k V_k(e_{k|k-1}). \end{cases} \tag{3.80}$$

Secondly, substituting (3.75), (3.77) into (3.80) leads to

$$E[w_k] \leq \frac{a_k\phi_k(1-\varsigma)|e_{k|k-1}|}{(1+\phi_k)(2a_k\phi_k+1-a_k)}, \tag{3.81}$$

$$E[w_k^2] \leq \frac{\phi_k(M-1)r_k(1-\varsigma)|e_{k|k-1}|^2 - 4.92M\hat{z}_{k|k-1}}{2M(1+\phi_k)^2}. \tag{3.82}$$

Thirdly, let

$$\frac{\phi_k(M-1)r_k(1-\varsigma)|e_{k|k-1}|^2}{2M(1+\phi_k)^2} \geq \frac{4.92\hat{z}_{k|k-1}}{(1+\phi_k)^2}, \tag{3.83}$$

and we thus have

$$|e_{k|k-1}| \geq \sqrt{\frac{9.84M\hat{z}_{k|k-1}}{\phi_k(M-1)r_k(1-\varsigma)}} \triangleq \tilde{\epsilon}, \tag{3.84}$$

$$E[w_k^2] \leq \frac{2.46\hat{z}_{k|k-1}}{(1+\phi_k)^2} \triangleq \sigma_k. \tag{3.85}$$

The rational behind can be interpreted as follows: (1) the right term of (3.82) cannot be less than zero and (2) there always exists the measurement uncertainty in the system. Consequently, the impact of tag population change plus the measurement uncertainty should equal in order of magnitude that of only measurement uncertainty, which can be achieved by establishing $E[w_k^2] \leq K_k^2 E[u_k^2]$ and (3.83) with reference to (3.76) and (3.77).

However, since a_k' and a_k are unknown a priori, we thus need to transform the right hand side of (3.81) to a computable form. From (3.63), we get $\frac{1}{a_k} - 1 \leq \frac{\varsigma}{1+\varsigma}$ such that it holds for the right hand side of (3.81) that $\frac{a_k \phi_k (1-\varsigma)|e_{k|k-1}|}{3(1+\phi_k)(2a_k \phi_k+1-a_k)} \geq \frac{\phi_k (1-\varsigma)\tilde{\epsilon}}{3(1+\phi_k)\left(2\phi_k + \frac{\varsigma}{1+\varsigma}\right)}$.

Finally, let

$$E[w_k] \leq \frac{\phi_k (1-\varsigma)\tilde{\epsilon}}{3(1+\phi_k)\left(2\phi_k + \frac{\varsigma}{1+\varsigma}\right)} \triangleq \lambda_k, \tag{3.86}$$

we can establish (3.79) and thus get that $E[e_{k|k-1}^2] \to 0$ at an exponential rate when $k \to \infty$.

- *Region 2:* $\sqrt{\frac{9.84\hat{z}_{k|k-1}}{\phi_k r_k(1-\varsigma)}} \leq |e_{k|k-1}| < \sqrt{\frac{9.84 M \hat{z}_{k|k-1}}{\phi_k (M-1)r_k(1-\varsigma)}}$. Given $\tilde{\epsilon}$, λ_k and σ_k as in *Region 1*, in this case, we have $-(1-\varsigma)\alpha_k V_k(e_{k|k-1}) + d_k |e_{k|k-1}| E[w_k] + \xi_k^* \leq 0$. It then follows from Lemma 3.2 that

$$E[e_{k|k-1}^2] \leq \frac{\overline{p}_k}{Q_0} E[e_{1|0}^2] \prod_{i=1}^{k-1}(1-\alpha_i^*) + \overline{p}_k \sum_{i=1}^{k-2} \tau_{k-i-1} \prod_{j=1}^{i}(1-\alpha_{k-j}^*). \tag{3.87}$$

Hence, when $k \to \infty$, $E[e_{k|k-1}^2]$ converges exponentially to $\overline{p}_k \sum_{i=1}^{k-2} \tau_{k-i-1} \cdot \prod_{j=1}^{i}(1-\alpha_{k-j}^*) \sim \Theta(\hat{z}_{k|k-1})$ and it thus holds that $\frac{E[e_{k|k-1}]}{z_k} = \Theta(\frac{1}{\sqrt{z_k}}) \to 0$ for $z_k \to \infty$.

- *Region 3:* $0 \leq |e_{k|k-1}| < \sqrt{\frac{9.84\hat{z}_{k|k-1}}{\phi_k r_k(1-\varsigma)}}$. The circumstances in this region are very complicated due to $E[w_k]$ and $E[w_k^2]$, we here thus just consider the worst case that $E[w_k] = \lambda_k$ and $E[w_k^2] = \sigma_k$. Consequently, we have $-(1-\varsigma)\alpha_k V_k(e_{k|k-1}) + d_k |e_{k|k-1}| E[w_k] + \xi_k^* > 0$, and it then follows from Lemma 3.2 that

$$E[e_{k|k-1}^2] \leq \frac{\overline{p}_k}{Q_0} E[e_{1|0}^2] \prod_{i=1}^{k-1}(1-\alpha_i^*) + \overline{p}_k \sum_{i=1}^{k-2} \tau_{k-i-1} \prod_{j=1}^{i}(1-\alpha_{k-j}^*). \tag{3.88}$$

Hence, when $k \rightarrow \infty$, $E[e_{k|k-1}^2]$ converges at exponential rate to $\overline{P_k} \sum_{i=1}^{k-2} \tau_{k-i-1} \prod_{j=1}^{i} (1 - \alpha_{k-j}^*) \sim \Theta(\hat{z}_{k|k-1})$, and thus $\frac{E[e_{k|k-1}]}{z_k} \leq \Theta(\frac{1}{\sqrt{z_k}}) \rightarrow 0$ for $z_k \rightarrow \infty$.

Note that for the case that $E[w_k] < \lambda_k$ and $E[w_k^2] < \sigma_k$, the range of *Region 3* will shrink and the range of *Region 2* will largen.

Integrating the above three regions, we can get the similar results on the convergence of the expected estimation error $E[e_{k|k-1}]$ as in the static case.

3.8   Discussion

This section discusses the application of the proposed algorithm to the unreliable channel and multi-reader scenario.

Error-Prone Channel   The unreliable channel may corrupt a would-be idle slot into a busy slot and vice versa. We consider the random error model as [30]. Let t_0 and t_1 be the false negative rate that a would-be empty slot turns into a busy slot and the false positive rate, respectively. Each parameter without error rate is marked with a superscript $*$ to define its counterpart with error rate t_0 and t_1. Then, we have

$$p^*(Z_k) = t_1 + (1 - t_0 - t_1)p(Z_k) \tag{3.89}$$

$$Var^*[u_k] = (1 - t_0 - t_1)^2 Var[u_k], \tag{3.90}$$

and thus get the new Kalman gain K_k^* as

$$K_k^* = \frac{1}{(1 - t_0 - t_1)} K_k. \tag{3.91}$$

It is noted that the ideal channel condition is equivalent to the special case where $t_0 = t_1 = 0$. When the channel is totally random, i.e., $t_0 = t_1 = 0.5$, the noisy will overwhelm the measurement and estimation. Nevertheless, if $t_0, t_1 \in (0, 0.5)$, updating the analysis with the new equations, we find that Theorem 3.1 and 3.2 still holds under the same conditions, meaning that the communication error can be compensated successfully.

Multi-Reader Case   In multi-reader scenarios, we leverage the same approach as [31]. The main idea is that a back-end server can be used to synchronize all readers such that the RFID system with multiple readers operates as the single-reader case. Specially, the back-end server calculates all the parameters and sends them to all readers such that they broadcast the same parameters to the tags. Subsequently, each reader sends its bitmap to the back-end server. Then the back-end server applies OR operator on all bitmaps, which eliminates the impact of the duplicate readings of tags in the overlapped interrogation region.

3.9   Numerical Analysis

In this section, we conduct extensive simulations to evaluate the performance of the proposed tag population estimation algorithms by focusing on the relative estimation error denoted as $REE_k = \left| \frac{z_k - \hat{z}_{k|k-1}}{z_k} \right|$. Specifically, we simulate in sequence both static and dynamic RFID systems where the initial tag population are $z_0 = 10^4$ with the following parameters: $q = 0.1$, $P_{0|0} = 1$, $J = 3$, $\theta = 4$ and $\Upsilon = 0.5$ with reference to (3.35) and (3.36), $L = 1500$, $\underline{\phi} = 0.25$ and $\overline{\phi} = 100$ such that (3.64) always holds. Since the proposed algorithms do not require collision detection, we set a slot to 0.4ms as in the EPCglobal C1G2 standard [12]. We will discuss the effect of $\underline{\phi}$ and $\overline{\phi}$ on the performance in next section.

3.9.1   Algorithm Verification

In the subsection, we show the impact of $\underline{\phi}$ and $\overline{\phi}$ on the system performance. To that end, with $REE_0 = 0.5$, we keep $z_k = 10^4$ in static scenario while the tag population varies in order of magnitude from $\sqrt{\hat{z}_{k|k-1}}$ to $0.4\hat{z}_{k|k-1}$ in different patterns in dynamic scenario. Specifically, we set $\overline{\phi} = 100$ while varying $\underline{\phi} = 0.25, 0.5, 1$ in Figs. 3.2, and 3.3, and fix $\underline{\phi} = 0.25$ with varying $\overline{\phi} = 1, 10, 100$ in Figs. 3.4, and 3.5. As shown in the figures, a smaller $\underline{\phi}$ leads to rapider convergence rate while the bigger $\overline{\phi}$, the smaller the deviation. Thus, we choose $\underline{\phi} = 0.25$ and $\overline{\phi} = 100$ in the rest of the simulation.

Fig. 3.2 Static:$\overline{\phi} = 100$

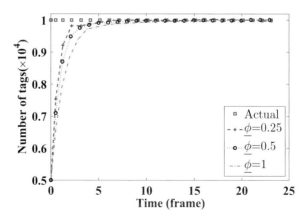

Fig. 3.3 Dynamic:$\overline{\phi} = 100$

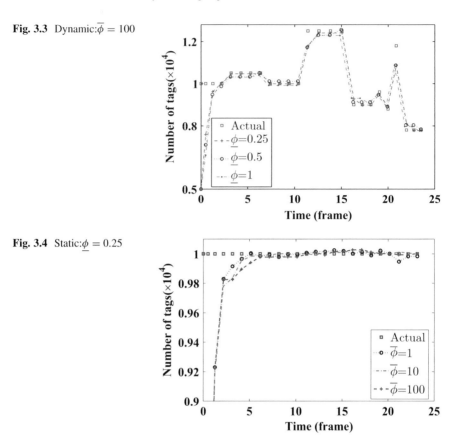

Fig. 3.4 Static:$\underline{\phi} = 0.25$

Moreover, we make the following observations. First, as derived in Theorem 3.2, the estimation is stable and accurate facing to a relative small population change, i.e., around the order of magnitude $\sqrt{\hat{z}_{k|k-1}}$. Second, the proposed scheme also functions nicely even when the estimation error is as high as $0.4\hat{z}_{k|k-1}$ tags as shown in Figs. 3.3 and 3.5. This is due to the CUSUM-based change detection which detects state changes promptly such that a small value is set for ϕ_k, leading to rapid convergence rate.

3.9.2 Algorithm Performance

In this section, we evaluate the performance of the proposed EKF-based estimator, referred to as EEKF here, in comparison with [7] in static scenario and with [19] in dynamic scenario.

Fig. 3.5 Dynamic:$\phi = 0.25$

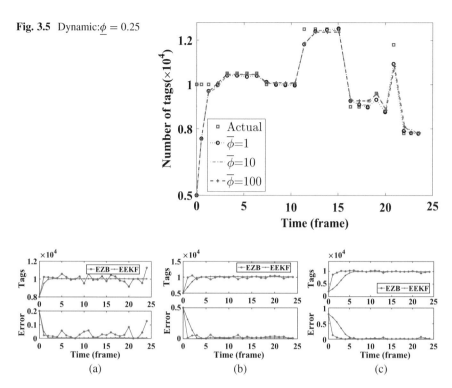

Fig. 3.6 Algorithm performance under different initial estimation errors. (**a**) $REE_0 = 0.2$. (**b**) $REE_0 = 0.5$. (**c**) $REE_0 = 0.8$

3.9.2.1 Static System ($z_k = 10^4$)

We evaluate the performance by varying initial relative error as

- $REE_0 = \frac{z_0 - \hat{z}_{0|0}}{z_0} = 0.8$ means a large initial estimation error.
- $REE_0 = 0.5$ means a medium initial estimation error.
- $REE_0 = 0.2$ implies a small initial estimation error and satisfies (3.66) with $0.5 \leq \varsigma < 1$.

The purpose of the first two cases is to investigate the effectiveness of the estimation in relative large initial estimation errors while the third case intends to verify the analytical results $\frac{\hat{z}_{0|0}}{z_0} > 0.79$ as shown in (3.66). Note that EZB uses the optimal persistence probability [7] that needs to know coarse range of tag size. And estimation time of EZB increases with the width of the range. Figure 3.6 illustrates the estimation processes with different initial estimation errors. As shown in the figures, the estimation $\hat{z}_{k|k-1}$ converges towards the actual number of tags within very short time in all cases, despite the initial estimation error. It is worth noticing that EZB suffers the significant deviation though it is faster than EEKF.

3.9.2.2 Dynamic System

Table 3.2 Execution time

Algo.	Variation of tag population								
	12500	6737	9364	7049	8616	11143	13385	8713	10761
JREP	3.28	3.28	3.28	3.28	3.28	3.28	3.28	3.28	3.28
EEKF	1.2	1.2	1.2	0.6	0.6	1.8	0.6	0.6	1.8

In this subsection, we evaluate the performance of EEKF for dynamic systems by comparing with the start-of-the-art solution JREP [19] in terms of execution time to achieve the required accuracy. To that end, we refer to the simulation setting in [19]. Specifically, the initial estimation error is 10%. The tag population size changes by following the normal distribution with the mean of 10,000 and the variance of 2000^2 and the accuracy requirement is 95%. By taking 9 samplings, we obtain the results as listed in Table 3.2. As shown in Table 3.2, EEKF is more time-efficient than JREP. This is because the persistence probability in JREP is set to optimise the power-of-two frame size, which increases the variance of the number of empty slots and leads to the performance degradation. In contrast, EEKF can minimise this variance while promptly detecting the tag population changes.

3.10 Conclusion

In this chapter, we have addressed the problem of tag estimation in dynamic RFID systems and designed a generic framework of stable and accurate tag population estimation schemes based on Kalman filter. Technically, we leveraged the techniques in extended Kalman filter (EKF) and cumulative sum control chart (CUSUM) to estimate tag population for both static and dynamic systems. By employing Lyapunov drift analysis, we mathematically characterised the performance of the proposed framework in terms of estimation accuracy and convergence speed by deriving the closed-form conditions on the design parameters under which our scheme can stabilise around the real population size with bounded relative estimation error that tends to zero within exponential convergence rate.

References

1. RFID Journal, DoD releases final RFID policy. [Online]
2. RFID Journal, DoD reaffirms its RFID goals. [Online]
3. C.-H. Lee, C.-W. Chung, Efficient storage scheme and query processing for supply chain management using RFID, in *ACM SIGMOD* (ACM, New York, 2008), pp. 291–302
4. L.M. Ni, D. Zhang, M.R. Souryal, RFID-based localization and tracking technologies. IEEE Wirel. Commun. **18**(2), 45–51 (2011)

5. P. Yang, W. Wu, M. Moniri, C.C. Chibelushi, Efficient object localization using sparsely distributed passive RFID tags. IEEE Trans. Ind. Electron. **60**(12), 5914–5924 (2013)
6. RFID Journal, Wal-Mart begins RFID process changes. [Online]
7. M. Kodialam, T. Nandagopal, W.C. Lau, Anonymous tracking using RFID tags, in *IEEE INFOCOM* (IEEE, Piscataway, 2007), pp. 1217–1225
8. T. Li, S. Wu, S. Chen, M. Yang, Energy efficient algorithms for the RFID estimation problem, in *IEEE INFOCOM* (IEEE, Piscataway, 2010), pp. 1–9
9. C. Qian, H. Ngan, Y. Liu, L. M. Ni, Cardinality estimation for large-scale RFID systems. IEEE Trans. Parallel Distrib. Syst. **22**(9), 1441–1454 (2011)
10. M. Shahzad, A.X. Liu, Every bit counts: fast and scalable RFID estimation, in *ACM Mobicom* (2012), pp. 365–376
11. Y. Zheng, M. Li, Zoe: fast cardinality estimation for large-scale RFID systems, in *IEEE INFOCOM* (IEEE, Piscataway, 2013), pp. 908–916
12. EPCglobal Inc., Radio-frequency identity protocols class-1 generation-2 UHF RFID protocol for communications at 860 mhz - 960 mhz version 1.0.9 [Online]
13. Y. Song, J.W. Grizzle, The extended Kalman filter as a local asymptotic observer for nonlinear discrete-time systems, in *American Control Conference* (IEEE, Piscataway, 1992), pp. 3365–3369
14. M. Kodialam, T. Nandagopal, Fast and reliable estimation schemes in RFID systems, in *ACM Mobicom* (ACM, New York, 2006), pp. 322–333
15. H. Han, B. Sheng, C.C. Tan, Q. Li, W. Mao, S. Lu, Counting RFID tags efficiently and anonymously, in *IEEE INFOCOM* (IEEE, Piscataway, 2010), pp. 1–9
16. V. Sarangan, M. Devarapalli, S. Radhakrishnan, A framework for fast RFID tag reading in static and mobile environments. Comput. Netw. **52**(5), 1058–1073 (2008)
17. L. Xie, B. Sheng, C.C. Tan, H. Han, Q. Li, D. Chen, Efficient tag identification in mobile RFID systems, in *IEEE INFOCOM* (IEEE, Piscataway, 2010), pp. 1–9
18. Q. Xiao, B. Xiao, S. Chen, Differential estimation in dynamic RFID systems, in *IEEE INFOCOM* (IEEE, Piscataway, 2013), pp. 295–299
19. Q. Xiao, M. Chen, S. Chen, Y. Zhou, Temporally or spatially dispersed joint RFID estimation using snapshots of variable lengths, in *ACM MobiHoc* (ACM, New York, 2015), pp. 247–256
20. T. Morozan, Boundedness properties for stochastic systems, in *Stability of Stochastic Dynamical Systems* (Springer, Berlin, 1972), pp. 21–34
21. T.-J. Tarn, Y. Rasis, Observers for nonlinear stochastic systems. IEEE Trans. Autom. Control **21**(4), 441–448 (1976)
22. K. Reif, S. Günther, E. Yaz Sr., R. Unbehauen, Stochastic stability of the discrete-time extended Kalman filter. IEEE Trans. Autom. Control **44**(4), 714–728 (1999)
23. M.B. Rhudy, Y. Gu, Online stochastic convergence analysis of the Kalman filter. Int. J. Stoch. Anal. **2013**, 240295 (2013)
24. K. Finkenzelle, *RFID Handbook: Radio Frequency Identification Fundamentals and Applications* (Wiley, Chichester, 2000)
25. V.F. Kolchin, B.A. Sevastyanov, V.P. Chistyakov, *Random Allocation* (Wiley, New York, 1978)
26. F. Gustafsson, F. Gustafsson, *Adaptive Filtering and Change Detection* (Wiley, New York, 2000)
27. E. Brodsky, B.S. Darkhovsky, *Nonparametric Methods in Change Point Problems* (Springer Science & Business Media, New York, 1993)
28. M. Basseville, I.V. Nikiforov, et al., *Detection of Abrupt Changes: Theory and Application* (Prentice Hall, Englewood Cliffs, 1993)
29. F. Spiring, Introduction to statistical quality control. Technometrics **49**(1), 108–109 (2007)
30. M. Chen, W. Luo, Z. Mo, S. Chen, Y. Fang, An efficient tag search protocol in large-scale RFID systems with noisy channel, in *IEEE/ACM TON* (2015)
31. M. Shahzad, A.X. Liu, Expecting the unexpected: fast and reliable detection of missing RFID tags in the wild, in *IEEE INFOCOM* (2015), pp. 1939–1947

Chapter 4
Finding Needles in a Haystack: Missing Tag Detection in Large RFID Systems

Chapter Roadmap The rest of this chapter is organised as follows. Section 4.1 explains the motivation of studying missing tag detection problem in RFID systems with the presence of unexpected tags and summarizes the contributions. Section 4.2 gives a brief review of existing missing tag detection and identification algorithms. Section 4.3 formulates the missing tag detection problem in RFID systems with the presence of unexpected tags. Section 4.4 details the missing tag detection protocol. Section 4.5 conducts theoretical performance analysis. Section 4.6 discusses the tag population estimation and its utility to rough missing tag detection. Section 4.7 shows the experimental results. Section 4.8 gives the summary.

4.1 Introduction

4.1.1 Motivation and Problem Statement

According to the statistics presented in [1], inventory shrinkage, a combination of shoplifting, internal theft, administrative and paperwork error, and vendor fraud, resulted in 44 billion dollars in loss for retailers in 2014. Fortunately, RFID technology can be used to reduce the cost by monitoring products for its low cost and non-line-of-sight communication pattern. Obviously, the first step in the application of loss prevention is to determine whether there is any missing tag. Hence, quickly finding out the missing tag event is of practical importance.

The presence of unexpected tags, however, prolongs the detection time and even leads to miss detection. Here, we present two examples to motivate the presence of unexpected tags in realistic scenarios.

- *Example 1.* Consider a retail store with expensive goods and a much larger amount of inexpensive goods, and an RFID system is deployed to monitor the goods. Because of the higher value of expensive products, they are expected to

© Springer International Publishing AG, part of Springer Nature 2019   77
J. Yu, L. Chen, *Tag Counting and Monitoring in Large-Scale RFID Systems*,
https://doi.org/10.1007/978-3-319-91992-8_4

Fig. 4.1 Missing tag
detection with the presence of
unexpected tags

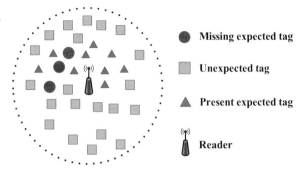

be detected more frequently, but the tags of inexpensive goods also response the
interrogation of readers, which influences the decision of readers.

* *Example 2.* Consider a large warehouse rented to multiple companies where the
products of the same company may be placed in different zones according to
their individual categories, such as child food and adult food, chilled food and
ambient food. When detecting the tags identifying products from one company,
readers also receive the feedbacks from the tags of other companies.

In both examples, how to effectively reduce the impact of unexpected tags is of
critical importance in missing tag detection. In this chapter, we consider a scenario,
as depicted in Fig. 4.1, where each product is affixed by an RFID tag. The reader
stores the IDs of expected tags. The problem we address is how to detect missing
expected tags in the presence of a large number of unexpected tags in the RFID
systems in a reliable and time-efficient way.

4.1.2   Prior Art and Limitation

Prior related work can be classified into three categories from the perspective of
detecting missing tags: missing tag detection protocols, tag identification protocols,
and tag estimation protocols.

There are two types of missing tag detection protocols: probabilistic [2–5] and
deterministic [6–8]. The probabilistic protocols find out a missing tag event with a
certain required probability if the number of missing tags exceeds a given threshold,
thus they are more time-efficient but return weaker results in comparison with the
deterministic protocols that report all IDs of the missing tags. Actually, they can be
used together such that a probabilistic protocol is executed in the first phase as an
alarm that reports the absence of tags and then a deterministic protocol is executed to
report IDs of missing tags. Unfortunately, all missing tag detection protocols except
RUN [5] work on the hypothesis of a perfect environment without unexpected tags
and thus fail to effectively detect missing tags in the presence of unexpected tags.
Although RUN [5] is tailored for missing tag detection in the RFID systems with

unexpected tags, all unexpected tags may always participate in the interrogation, which leads to the significant degradation of the performance when the unexpected tag population size scales.

Tag identification protocols [9–12] can identify all tags in the interrogation region. To detect missing tags, tag identification protocols can be executed to obtain the IDs of the tags present in the population and then the missing tags can be found out by comparing the collected IDs with those recorded in the database. However, they are usually time-consuming [6] and fail to work when it is not allowed to read the IDs of tags due to privacy concern.

Tag estimation protocols [13–16] are used to estimate the number of tags in the interrogation region. If many expected tags are absent in RFID systems without unexpected tags, a missing tag event may be detected by comparing the estimation and the number of expected tags stored in the database. However, the estimation error may be misinterpreted as missing tags and cause detection error, especially when there are only a few missing tags. Moreover, the estimation protocol cannot handle the case with a large number of unexpected tags.

4.1.3 Proposed Solution and Main Contributions

Motivated by the detrimental effects of unexpected tags on the performance of missing tag detection, we devise a reliable and time-efficient protocol named Bloom filter-based missing tag detection protocol (BMTD). Specifically, BMTD consists of two phases, each consisting of a number of rounds.

- In each round of the first phase, the reader fist constructs a Bloom filter by mapping all the expected tag IDs into it such that each tag has multiple representative bits. Then the constructed Bloom filter is broadcasted to all tags. If at least one representative bit of a tag is '0's, it finds itself unexpected and will not participate in the rest of BMTD. Thus, the number of active unexpected tags is considerably reduced.
- Subsequently, in each round of the second phase, the reader constructs a Bloom filter by aggregating the feedbacks from the remaining tags and uses it to check whether any expected tag is absent from the population.

The major contributions of this chapter can be articulated as follows. First, we propose a new solution for the important and challenging problem of missing tag detection in the presence of a large number of unexpected tags by employing Bloom filter to filter out the unexpected tags and then detect the missing tags. Second, we perform the theoretical analysis for determining the optimal parameters used in BMTD that minimize the detection time and also meet the required reliability. Third, we perform extensive simulations to evaluate the performance of BMTD. The results show that BMTD significantly outperforms the state-of-the-art solutions.

4.2 Related Work

Extensive research efforts have been devoted to detecting missing tags by using probabilistic method [2–5] and deterministic method [6–8]. Next, we briefly review the existing solutions of missing tag detection problem and introduce Bloom filter.

4.2.1 Probabilistic Protocols

The objective of probabilistic protocols is to detect a missing tag event with a predefined probability. Tan et al. initiate the study of probabilistic detection and propose a solution called TRP in [2]. TRP can detect a missing tag event by comparing the pre-computed slots with those picked by the tags in the population. Different from our BMTD, TRP does not take into account the negative impact of unexpected tags. Follow-up works [3, 4] employ multiple seeds to increase the probability of the singleton slot. Same to TRP, they are required to know all the tags in the population. The latest probabilistic protocol called RUN is proposed in [5]. The difference with previous works lies in that RUN considers the influence of unexpected tags and can work in the environment with unexpected tags. However, RUN does not eliminate the interference of unexpected tags fundamentally such that the false positive probability does not decrease with respect to the unexpected tag population size, which shackles the detection efficiency especially in the presence of a large number of unexpected tags. In addition, the first frame length is set to the double of the cardinality of the expected tag set in RUN, which is not established by theoretical analysis and leads to the failure of estimation method in RUN when the number of the unexpected tags is far larger than that of the expected tags.

4.2.2 Deterministic Protocols

The objective of deterministic protocols is to exactly identify which tags are absent. Li et al. develop a series of protocols in [6] which intend to reduce the radio collision and identify a tag not in the ID level but in the bit level. Subsequently, Zhang et al. propose another series of determine protocols in [7] of which the main idea is to store the bitmap of tag responses in all rounds and compare them to determine the present and absent tags. But how to configure the protocol parameters is not theoretically analyzed. More recently, Liu et al. [8] enhance the work by reconciling both 2-collision and 3-collision slots and filtering the empty slots such that the time efficiency can be improved. None of existing deterministic protocols, however, have been designed to work in the chaotic environment with unexpected tags.

4.2.3   Bloom Filter

A Bloom filter is a randomized data structure that is originally from database contexts [17, 18] and is used to records the members of a set but has attracted much research attention in networking applications [19]. Specifically, given a set $A = \{a_1, a_2, \cdots, a_n\}$, Bloom filter operates as follows:

Initialization Let a bit array BF be the Bloom filter and the length of BF be f, we initialize BF with zero array. Then the filter is incrementally built by inserting items of A by setting certain bits of BF to 1.

Insertion To insert an arbitrary item $a_i \in A$, we first need to feed a_i to k independent hash functions $h_1, h_2, \cdots, h_k$ to retrieve k values: $h_v(a_i) \mod f$ for $1 \leq v \leq k$, which directs to k positions in BF. Insertion of a_i is then achieved by setting the bits in these k positions to 1.

Query To determine whether an item b belongs to A, we can check if b has been inserted into the Bloom filter BF. Achieving this requires b to be inserted by the same hash functions and then we can check every bit $BF[h_v(b) \mod f]$ for $1 \leq v \leq k$. If all of k bits are set to 1, the Bloom filter asserts $b \in A$; otherwise, $b \notin A$.

4.3   System Model and Problem Formulation

4.3.1   System Model

Consider a large RFID system consisting of a single RFID reader and a large number of RFID tags. The reader broadcasts the commands and collects the feedbacks from the tags. In the RFID system, the tags can be either battery-powered active ones or lightweight passive ones that are energized by radio waves emitted from the reader. In this chapter, we first take account of the single-reader case and then extend the proposed protocol to the multi-reader case.

The communications between the readers and the tags follow the *Listen-before-talk* mechanism [20].During the communications, the tag-to-reader transmission rate and the reader-to-tag transmission rate may differ with each other and are subject to the environment. In practice, the former can be either $40 \sim 640$ kb/s in the FM0 encoding format or $5 \sim 320$ kb/s in the modulated subcarrier encoding format, while the later is normally about $26.7 \sim 128$ kb/s [21].

4.3.2   Problem Formulation

In the considered RFID system, we use $\mathbb{E}$ to denote the set of IDs of the expected tags which are expected to be present in a population and target tags to be monitored. In the RFID system, we assume that an unknown number of tags, m, out of these $|\mathbb{E}|$

Table 4.1 Main notations

Symbols	Descriptions
$\mathbb{E}$	Set of target tags that need to be monitored
$\mathbb{E}_r$	Tags that are actually present in the population
$\mathbb{U}$	Set of unexpected tags
α	Required detection reliability
m	Number of missing expected tags
M	Threshold to detect missing tags
P_{sys}	Prob. of detecting a missing event in BMTD
J	Number of rounds in Phase 1
l_j	Length of the j-th frame of Phase 1
k_j	Number of hash functions in the j-th frame of Phase 1
s_j	Random seed used in the j-th frame of Phase 1
$\mathbb{U}_r$	Set of remaining active unexpected tags after Phase 1
N^*	Number of remaining active tags after Phase 1
$P_{1,j}$	False positive rate in the j-th frame of Phase 1
T_1	Time cost of Phase 1
W	Number of rounds in Phase 2
f_w	Length of the w-th frame of Phase 2
R_w	Number of hash functions in the w-th frame of Phase 2
d_w	Random seed used in the w-th frame of Phase 2
$P_{2,w}$	False positive rate in the w-th frame of Phase 2
T_2	Time taken to execute W rounds in Phase 2
T	Theoretical execution time
q	Prob. of detect a missing tag in a given slot of Phase 2
Z	Random variable for slot of the first detection
$E[T_D]$	Expected detection time of BMTD

tags are missing. Note that $|\cdot|$ stands for the cardinality of a set. Denote by $\mathbb{E}_r$ the set of IDs of the remaining $|\mathbb{E}| - m$ tags that are actually present in the population. Let $\mathbb{U}$ be the set of IDs of unexpected tags within the interrogation region of the reader which does not need to be monitored. The reader may neither knows exactly the IDs of unexpected tags nor does it know the cardinality of $\mathbb{U}$.

Let M be a threshold on the number of missing expected tags. We use P_{sys} to denote the probability that the reader can detect a missing event. The optimum missing tag detection problem is formally defined as follows.

Definition 4.1 (Optimum Missing Tag Detection Problem) Given $|\mathbb{U}|$ unexpected tags where both $|\mathbb{U}|$ and the IDs of tags in $\mathbb{U}$ are unknown, the optimum missing tag detection problem is to devise a protocol of minimum execution time capable of detecting a missing event with probability $P_{sys} \geq \alpha$ if $m \geq M$, where α is the system requirement on the detection reliability.

Table 4.1 summaries the main notations used in this chapter.

4.4 Bloom Filter-Based Missing Tag Detection Protocol

4.4.1 Design Rational and Protocol Overview

To improve the time efficiency of detecting missing tags in the presence of a large number of unexpected tags in the population, we limit the interference of unexpected tags in our protocol. To achieve this goal, we employ a powerful technique called *Bloom filter* which is a space-efficient probabilistic data structure for representing a set and supporting set membership queries [17] to rule out the unexpected tags in the set $\mathbb{U}$, which efficiently reduces their interference and thus the overall execution time. Following this idea, we propose a *Bloom filter-based Missing Tag Detection protocol* (BMTD), by which Bloom filters are sequentially constructed by the reader and by the feedbacks from the active tags in the RFID system.

The BMTD consists of two phases: (1) the unexpected tag deactivation phase and (2) the missing tag detection phase.

- The first phase is divided into J rounds where the reader constructs J Bloom filters by mapping the recorded IDs in the reader to deactivate the unexpected tags after identifying them.
- The second phase is divided into W rounds. The reader constructs W Bloom filters according to the responses of the remaining active tags and uses the Bloom filters to detect any missing event. Our protocol either detects a missing event or reports no missing event if the reader does not detect a missing event after W rounds.

We elaborate the design of the BMTD in the rest of this section.

4.4.2 Phase 1: Unexpected Tag Deactivation

In Phase 1, we use Bloom filters to reduce the number of active unexpected tags. Specifically, in the j-th round of Phase 1 ($j = 1, 2, \ldots, J$), the reader first constructs a Bloom filtering vector by mapping the expected tags in set $\mathbb{U}$ into an l_j-bit array using k_j hash functions with random seed s_j. Here, we denote the l_j-bit Bloom filter vector as $BF_{1,j}(\mathbb{E})$. How the values of l_j, k_j are chosen and how J is calculated are analysed in Sect. 4.5 on parameter optimisation.

Then, the reader broadcasts the l_j-bit Bloom filtering vector, k_j and s_j to all tags. Upon receiving $BF_{1,j}(\mathbb{E})$, k_j, and s_j, each tag maps its ID to k_j bits pseudo-randomly at positions $h_1(ID), h_2(ID), \cdots, h_{k_j}(ID)$, and checks the corresponding positions in $BF_{1,j}(\mathbb{E})$. If all of k_j bits are 1, then the tag regards itself expected by the reader. If any of k_j bits is 0, the tag regards that it is unexpected and then remains silent in the rest of the time.

Let $\mathbb{U}_j$ denote the set of the remaining active unexpected tags after the j-th round of Phase 1, and let $\mathbb{U}_j \cap BF_{1,j}(\mathbb{E})$ denote the set of unexpected tags that pass the membership test of $BF_{1,j}(\mathbb{E})$. Since the Bloom filter has no false negatives, the set of remaining active tags can be represented as $\mathbb{E}_r \cup \mathbb{U}_{j-1} \cap BF_{1,j}(\mathbb{E})$.

After J rounds when Phase 1 is terminated, the number of remaining active unexpected tags, termed as $|\mathbb{U}_r|$, is $|\mathbb{U}_J \cap BF_{1,J}(\mathbb{E})|$. The present tag population size can be written as $|\mathbb{E}_r \cup \mathbb{U}_r|$. Subsequently, the reader enters Phase 2.

4.4.3 Phase 2: Missing Tag Detection

In the second phase, we still employ Bloom filter to detect a missing tag event. Note that the parameters that the reader broadcasts in each round in Phase 2 except random seeds are identical. In the w-th round of Phase 2 ($w = 1, 2, \ldots, W$), the reader first broadcasts the parameters containing the Bloom filter size f_w, the number of hash functions R_w, and a new random seed d_w. How their values are chosen and how W is calculated are analysed in Sect. 4.5 on parameter optimisation.

After receiving the configuration parameters, each tag in the set $\mathbb{E}_r \cup \mathbb{U}_r$ selects R_w slots at the indexes $h_v(ID)$ ($1 \leq v \leq R_w$) in the frame of f_w slots and transmits a short response at each of the R_w corresponding slots. As a consequence, a Bloom filter is formed in the air by the responses from the remaining active tags. In each round, there are two types of slots: empty slots and nonempty slots.

According to the responses from the tags, the reader encodes an f_w-bit Bloom filter as follows: If the i-th slot is empty, the reader sets i-th bit of the f_w-bit vector to be '0', otherwise '1'. Consequently, a virtual Bloom filter is constructed using which the reader then performs membership test. Let $BF_{2,w}(\mathbb{E}_r \cup \mathbb{U}_r)$ denote the constructed Bloom filter in w-th round.

To perform membership test, the reader uses tag IDs from the expected tag set $\mathbb{E}$. Specifically, for each ID in $\mathbb{E}$, the reader maps it into R_w bits at positions $h_v(ID)$ ($1 \leq v \leq R_w$) in $BF_{2,w}(\mathbb{E}_r \cup \mathbb{U}_r)$. If all of them are '1's, then the tag is regarded as present. Otherwise, the tag is considered to be missing. If a missing event is detected in w-round, the reader terminates the protocol without executing the remaining rounds. Otherwise, the reader initiates a new round until the protocol runs W rounds. If the reader does not detect a missing event after W rounds, it reports no missing event, i.e., the number of missing tags m is less than the threshold M.

4.4.4 An Illustrative Example of BMTD

We present an illustrative example to show the execution of BMTD. Consider an RFID system with 4 tags. We assume that the reader needs to monitor tag 1 and tag 2 and thus knows their IDs, i.e., $\mathbb{E} = \{ID1, ID2\}$, but it is not aware of the presence of tag 3 and tag 4, who are unexpected, i.e., $\mathbb{U} = \{ID3, ID4\}$. In the example, tag 2 is missing from the population.

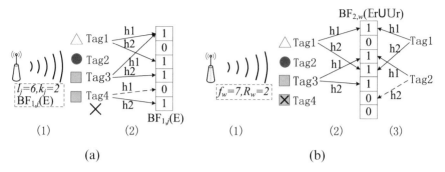

Fig. 4.2 Example illustrating BMTD. (**a**) Phase 1: unexpected tag deactivation, (**b**) Phase 2: missing tag detection

As shown in (1) of Fig. 4.2a, the reader first constructs a Bloom filter $BF_{1,j}(\mathbb{E})$ by mapping IDs in $\mathbb{E}$ and broadcasts a message containing $BF_{1,j}(\mathbb{E})$ and the values of k_j and l_j. Here we assume $J = 1$, $k_j = 2$ and $l_j = 6$. After receiving $BF_{1,j}(\mathbb{E})$, each tag checks if it is an expected tag. As shown in (2) of Fig. 4.2a, tag 1 finds itself expected due to the fact that both $h_1(ID1)$ and $h_2(ID1)$ are equal to 1. However, tag 4 realizes that it is unexpected for $h_1(ID4) = 0$ and deactivates itself. Different from tag 4, actually unexpected tag 3 passes the test and will participate in the rest of BMTD.

As depicted in (1) of Fig. 4.2b, after the first phase, the reader starts to detect missing tags by broadcasting parameters f_w and R_w. Here we assume $W = 1$, $R_w = 2$ and $f_w = 7$. By using f_w and R_w, tag 1 and tag 3 generate a Bloom filter vector, respectively, which is shown in (2) of Fig. 4.2b. Then they transmit following their individual Bloom filter vector. By sensing the channel, the reader can encode a Bloom filter and use it to check the IDs in $\mathbb{E}$ one by one. As shown in (3) of Fig. 4.2b, since the Bloom filter is constructed based on the responses of tag 1 and tag 3, tag 1 passes the test but tag 2 fails and is regarded as absent. Then the protocol reports a missing event.

4.5   Performance Optimisation and Parameter Tuning

In this section, we investigate how the parameters in the BMTD are configured to minimise the execution time while ensuring the performance requirement.

4.5.1   Tuning Parameters in Phase 1

According to the property of Bloom filter, false negatives are impossible. The false positive rate of the Bloom filter $BF_{1,j}(\mathbb{E})$ in the j-th round in Phase 1, defined as $P_{1,j}$, can be calculated as follows [17]:

$$P_{1,j} = \left[1 - \left(1 - \frac{1}{l_j}\right)^{|\mathbb{E}|k_j}\right]^{k_j} \approx (1 - e^{-|\mathbb{E}|k_j/l_j})^{k_j}. \tag{4.1}$$

By rearranging (4.1), we can express the Bloom filter size in the j-th round as

$$l_j = \frac{-|\mathbb{E}|k_j}{\ln(1 - P_{1,j}^{\frac{1}{k_j}})}. \tag{4.2}$$

The total time spent in this round can thus be calculated as $l_j * t_r$, where t_r denotes the per bit transmission time from reader to tags.

We denote C_j the cost to detect and deactivate an unexpected tag as follows:

$$C_j = \frac{l_j t_r}{|\mathbb{U}|(1 - P_{1,j})} = \frac{-t_r|\mathbb{E}|k_j}{|\mathbb{U}|(1 - P_{1,j})\ln(1 - P_{1,j}^{\frac{1}{k_j}})}. \tag{4.3}$$

From the expression of C_j, it can be noted that C_j represents the average time consumed to detect and deactive an unexpected tag in the j-th round. In our design we minimize C_j so as to achieve the optimal time-efficiency. To minimize C_j, we first compute the derivative of C_j with respect to k_j as follows:

$$\frac{dC_j}{dk_j} = \frac{|\mathbb{E}|t_r\left(P_{1,j}^{\frac{1}{k_j}}\ln P_{1,j} - k_j(1 - P_{1,j}^{\frac{1}{k_j}})\ln(1 - P_{1,j}^{\frac{1}{k_j}})\right)}{|\mathbb{U}|(1 - P_{1,j})k_j(1 - P_{1,j}^{\frac{1}{k_j}})\ln^2(1 - P_{1,j}^{\frac{1}{k_j}})}. \tag{4.4}$$

Furthermore, let $\frac{dC_j}{dk_j} = 0$, we can obtain

$$P_{1,j}^{\frac{1}{k_j}} = \frac{1}{2}, \tag{4.5}$$

and the unique minimiser $k_j^* = \frac{-\ln P_{1,j}}{\ln 2}$ as $\frac{dC_j}{dk_j} > 0$ when $k_j > \frac{-\ln p_{1,j}}{\ln 2}$, and $\frac{dC_j}{dk_j} < 0$ when $k_j < \frac{-\ln p_{1,j}}{\ln 2}$. Therefore, C_j reaches the minimum value when $P_{1,j}^{\frac{1}{k_j^*}} = \frac{1}{2}$. The optimum Bloom filter size, denoted as l_j^*, can be computed as

$$l_j^* = \frac{|\mathbb{E}|k_j^*}{\ln 2}. \tag{4.6}$$

The time spent in the j-th round can be computed as $\frac{|\mathbb{E}|t_r k_j^*}{\ln 2}$. Therefore, the total execution time of Phase 1, denoted as T_1, can be derived as

$$T_1 = \sum_{j=1}^{J} \frac{|\mathbb{E}| t_r k_j^*}{\ln 2}. \tag{4.7}$$

k_j^* ($1 \leq j \leq J$), as well as J, are set with the parameters in Phase 2 to minimize the global execution time, as analyzed in Sects. 4.5.3 and 4.5.4.

Let N^* be the number of tags still active after Phase 1 (i.e., J rounds), it holds that

$$N^* = |\mathbb{E}| - m + |\mathbb{U}_r|, \tag{4.8}$$

where $\mathbb{U}_r$ is the set of unexpected tags still active after Phase 1. Recall (4.5), the expectation of N^* can be derived as

$$E[N^*] = |\mathbb{E}| - m + |\mathbb{U}| \prod_{j=1}^{J} P_{1,j}$$

$$= |\mathbb{E}| - m + |\mathbb{U}| \left(\frac{1}{2}\right)^{\sum_{j=1}^{J} k_j^*}. \tag{4.9}$$

4.5.2   Tuning Parameters in Phase 2

Similar to Phase 1, the false positive rate of the w-th round in Phase 2, defined as $P_{2,w}$, can be calculated as

$$P_{2,w} = \left[1 - \left(1 - \frac{1}{f_w} \right)^{N^* R_w} \right]^{R_w} \approx (1 - e^{-N^* R_w / f_w})^{R_w}. \tag{4.10}$$

Therefore, the Bloom filter size is

$$f_w = \frac{-N^* R_w}{\ln(1 - P_{2,w}^{\frac{1}{R_w}})}.$$

Moreover, the probability that at least one missing tag can be detected in w-th round, denoted as $P_{d,w}$, can be computed as

$$P_{d,w} = 1 - P_{2,w}^m. \tag{4.11}$$

Following the analysis above, the probability P_{sys} that the reader is able to detect a missing event after at most W rounds in Phase 2, can thus be written as

$$P_{sys} = 1 - \prod_{w=1}^{W} (1 - P_{d,w}) = 1 - P_{2,w}^{mW}. \tag{4.12}$$

It follows from the system requirement that

$$P_{sys} = 1 - P_{2,w}^{mW} = \alpha. \tag{4.13}$$

As a result, we can obtain

$$f_w = \frac{-N^* R_w}{\ln(1 - (1 - \alpha)^{\frac{1}{mWR_w}})}. \tag{4.14}$$

In the following lemma, we derive the optimum frame size of the Bloom filter f_w which is broadcast by the reader in each round of Phase 2.

Lemma 4.1 *Let* $y \triangleq WR_w$, *the optimum Bloom filter frame size, denoted by f_w^*, that achieves the detection requirement while minimising the execution time of Phase 2, is as follows:*

$$f_w^* = \frac{-N^* R_w}{\ln(1 - (1 - \alpha)^{\frac{1}{my^*}})} \tag{4.15}$$

where $y^* = \frac{\ln(1-\alpha)}{m \ln \frac{1}{2}}$.

Proof Denote by f the total length of all W Bloom filters in the second phase, we thus have

$$f = \sum_{w=1}^{W} f_w = \frac{-N^* W R_w}{\ln(1 - (1 - \alpha)^{\frac{1}{mWR_w}})}. \tag{4.16}$$

It can be checked that f depends on the product of W and R_w which is the total number of hash functions used in Phase 2. To minimize the execution time, let $y \triangleq WR_w$, we first calculate the derivation of f with respect to y as follows:

$$\frac{df}{dy} = \frac{N^*(1 - \alpha)^{\frac{1}{my}} \ln(1 - \alpha)}{my(1 - (1 - \alpha)^{\frac{1}{my}}) \ln^2(1 - (1 - \alpha)^{\frac{1}{my}})} - \frac{N^*}{\ln(1 - (1 - \alpha)^{\frac{1}{my}})}.$$

Imposing $\frac{\mathrm{d}f}{\mathrm{d}y} = 0$ yields

$$y = \frac{\ln(1-\alpha)}{m \ln \frac{1}{2}}.$$

Moreover, when $y < \frac{\ln(1-\alpha)}{m \ln \frac{1}{2}}$, it holds that $\frac{\mathrm{d}f}{\mathrm{d}y} < 0$; when $y > \frac{\ln(1-\alpha)}{m \ln \frac{1}{2}}$, it holds that $\frac{\mathrm{d}f}{\mathrm{d}y} > 0$. Therefore, f achieves the minimum at $y^* = \frac{\ln(1-\alpha)}{m \ln \frac{1}{2}}$. The minimum of f_w, denoted by f_w^* can be computed by injecting $y = y^*$ into (4.14). The proof is thus completed. $\square$

Remark As the reader does not have prior knowledge on m, the number of missing tags, in the design of BMTD, we require that the detection performance requirement to be hold for any $m \geq M$. Hence, f_w^* and y^* are as follows:

$$f_w^* = \frac{-N^* R_w}{\ln(1 - (1-\alpha)^{\frac{1}{My^*}})}, \tag{4.17}$$

$$\text{where } y^* = \frac{\ln(1-\alpha)}{M \ln \frac{1}{2}}, \tag{4.18}$$

where we use $m = M$ in N^* and y^*, which is the hardest case. Since $N^* = |\mathbb{E}| - m + |\mathbb{U}_r|$, it can be checked that the detection probability P_{sys} is monotonically increasing and $P_{2,w}$ is monotonically decreasing with respect to the number of missing tags m, meaning that $m = M$ makes the detection hardest and any greater m will ease the hardness, it is thus reasonable to use $m = M$ in the rest of the analysis, because if the reader can detect a missing tag event with probability α when $m = M$, it will fulfill the detection with probability $P_{sys} > \alpha$ when $m > M$.

In addition, since y^* is the total number of hash functions used in Phase 2 and at least one round is executed so as to detect a missing event, y^* needs to be a positive integer. Therefore, we set $y^* = \lceil \frac{\ln(1-\alpha)}{M \ln \frac{1}{2}} \rceil$, which guarantees the required detection performance requirement. Note that R_w and W can be set as arbitrary positive integers.

Under the optimum parameter setting derived above, we can calculate the time needed to execute W rounds of Phase 2, denoted by T_2, as follows:

$$T_2 = \frac{-t_t N^* y^*}{\ln(1 - (1-\alpha)^{\frac{1}{My^*}})}, \tag{4.19}$$

where t_t is the time needed by the tags to transmit one bit to the reader. T_2 sets an upper-bound on the execution time of Phase 2.

4.5.3 Tuning k_j^* and J to Minimize Worst-Case Execution Time

In this subsection, we study how to set k_j^* and J to minimize the worst-case execution time, which corresponds to the experience of the execution time where no missing event is detected and hence all the W rounds in the second round need to be executed. We denote the worst-case execution time by T. In the following theorem, we derive the minimiser of $\mathbb{E}[T]$.

Theorem 4.1 *Denote $x \triangleq \sum_{j=1}^{J} k_j^*$, x need to be set to x^* as follows to minimise the worst-case execution time of the BMTD:*

$$
x^* = \begin{cases} 0 & |\mathbb{U}| \leq U_0 \\[2ex] \dfrac{\ln \frac{-t_r |\mathbb{E}| \ln(1-(1-\alpha^{\frac{1}{My^*}}))}{t_t y^* |\mathbb{U}| \ln^2 2}}{-\ln 2} & |\mathbb{U}| > U_0 \end{cases}, \tag{4.20}
$$

where $U_0 \triangleq \dfrac{|\mathbb{E}|t_r \ln(1-(1-\alpha)^{\frac{1}{My^*}})}{-t_t y^* \ln^2 2}$. That is, in regard to minimise the worst-case execution time, when the number of unexpected tags does not exceed a threshold U_0, Phase 1 is not executed, otherwise Phase 1 is executed with the parameters k_j^* and J set to $\sum_{j=1}^{J} k_j^* = x^*$.

Proof Recall the two phases of BMTD and (4.7), we can derive the expectation of T as follows:

$$
\mathbb{E}[T] = T_1 + T_2 = \sum_{j=1}^{J} \frac{|\mathbb{E}|t_r k_j^*}{\ln 2} + \frac{-t_t y^* E[N^*]}{\ln(1-(1-\alpha)^{\frac{1}{My^*}})}
$$

$$
= \frac{|\mathbb{E}|t_r}{\ln 2} \sum_{j=1}^{J} k_j^* + \frac{-t_t y^* \left(|\mathbb{E}| - M + |\mathbb{U}|(\frac{1}{2})^{\sum_{j=1}^{J} k_j} \right)}{\ln(1-(1-\alpha)^{\frac{1}{My^*}})}. \tag{4.21}
$$

From (4.21), it can be noted that $E[T]$ is a function of $x = \sum_{j=1}^{J} k_j^*$. We then calculate the optimum x^* that minimizes $E[T]$. To that end, we compute the derivation of $E[T]$ with respect to x:

$$
\frac{dE[T]}{dx} = \frac{|\mathbb{E}|t_r}{\ln 2} + \frac{t_t y^* |\mathbb{U}| \ln 2}{\ln(1-(1-\alpha)^{\frac{1}{My^*}})} (\frac{1}{2})^x. \tag{4.22}
$$

Since $(\frac{1}{2})^x \leq 1$, it thus holds for all $x \geq 0$ that $\frac{dE[T]}{dx} \geq 0$ if $\frac{|\mathbb{E}|t_r}{\ln 2} + \frac{t_t y^* |\mathbb{U}| \ln 2}{\ln(1-(1-\alpha)^{\frac{1}{My^*}})} \geq 0$, i.e.,

$$|\mathbb{U}| \le \frac{|\mathbb{E}|t_r \ln(1 - (1-\alpha)^{\frac{1}{My^*}})}{-t_t y^* \ln^2 2} = U_0. \tag{4.23}$$

It is worth noticing that $E[T]$ is a monotonic nondecreasing function in this case with respect to x, we thus set $x = 0$ to minimize the execution time, which means that if the number of unexpected tags is smaller than the threshold U_0, we should remove the Phase 1 and only execute Phase 2.

In contrast, if $|\mathbb{U}| > U_0$, $\frac{dE[T]}{dx}$ can be negative, zero, or positive. Setting $\frac{dE[T]}{dx} = 0$, the optimal value of x to minimise $E[T]$, defined as x^*, can be calculated as

$$x^* = \frac{\ln \frac{-t_r |\mathbb{E}| \ln(1-(1-\alpha)^{\frac{1}{My^*}})}{t_t y^* |\mathbb{U}| \ln^2 2}}{-\ln 2}.$$

$\square$

Remark Since x^* represents the total number of hash functions used in Phase 1, it needs to be a non-negative integer. Therefore, we set x^* either to its ceiling or floor integer depending on which one leads to a smaller $E[T]$. The parameters k_j^* and J are set such that $\sum_{j=1}^{J} k_j^* = x^*$.

4.5.4   Tuning k_j^* and J to Minimize Expected Detection Time

The parameters derived in Theorem 4.1 establish that the BMTD is able to detect a missing event with probability equal to or greater than the system requirement α after W rounds of Phase 2. However, in many practical scenarios, the missing event may be detected in the round $w < W$ when the algorithm can be terminated. In this subsection, we derive the parameter configuration (i.e., k_j^* and J) that minimises the expected detection time. To that end, we first calculate the probability that at least one of the missing tags can be detected for the first time in a given slot and use it to formulate the expectation of the missing event detection time.

Lemma 4.2 *The probability that a missing tag can be detected in a given slot of Phase 1, denoted by q, is as follows:*

$$q = \left(1 - (1 - (1-\alpha)^{\frac{1}{y^*M}})^{\frac{M}{N^*}}\right) \cdot \left(1 - (1-\alpha)^{\frac{1}{y^*M}}\right). \tag{4.24}$$

A loose lower-bound for q, denoted as q_{min}, can be established as follows:

$$q_{min} = \left(1 - \left(\frac{1}{2}\right)^{\frac{M}{|\mathbb{E}|-M+|\mathbb{U}|}}\right)(1 - (1-\alpha)^{\frac{1}{y^*M}}). \tag{4.25}$$

Proof A missing tag can be detected in a given slot only when at least one missing tag is hashed to this slot and no tag in $\mathbb{E}_r \cup \mathbb{U}_r$ selects the same location. Consider the hardest case for detecting a missing tag event, i.e., $m = M$, the probability that at least one missing tag maps to the given slot can be given by $\left(1 - (1 - \frac{1}{f_w^*})^{M R_w}\right)$. The probability that no tag in $\mathbb{E}_r \cup \mathbb{U}_r$ maps to that slot is equal to $(1 - \frac{1}{f_w^*})^{N^* R_w}$. Consequently, multiplying the former by the later leads to q, i.e.:

$$q = \left(1 - (1 - \frac{1}{f_w^*})^{M R_w}\right) \cdot (1 - \frac{1}{f_w^*})^{N^* R_w}$$

$$\approx (1 - e^{-\frac{M R_w}{f_w^*}}) \cdot e^{-\frac{N^* R_w}{f_w^*}}$$

$$= \left(1 - (1 - (1 - \alpha)^{\frac{1}{y^* M}})^{\frac{M}{N^*}}\right) \cdot (1 - (1 - \alpha)^{\frac{1}{y^* M}}).$$

We then derive the lower-bound q_{min}. To that end, noticing that q is negatively correlated with N^* which falls into the range $\left[|\mathbb{E}| - M, |\mathbb{E}| - M + |\mathbb{U}|\right]$, we have

$$q \geq \left(1 - (1 - (1 - \alpha)^{\frac{1}{y^* M}})^{\frac{M}{|\mathbb{E}| - M + |\mathbb{U}|}}\right) \cdot (1 - (1 - \alpha)^{\frac{1}{y^* M}}).$$

On the other hand, noticing that $y^* = \lceil \frac{\ln(1-\alpha)}{M \ln \frac{1}{2}} \rceil \geq \frac{\ln(1-\alpha)}{M \ln \frac{1}{2}}$, we have $q \geq q_{min} = \left(1 - (\frac{1}{2})^{\frac{M}{|\mathbb{E}| - M + |\mathbb{U}|}}\right)(1 - (1 - \alpha)^{\frac{1}{y^* M}})$. $\qquad\square$

After calculating q, we next derive the expected missing event detection time, denoted by $\mathbb{E}[T_D]$.

Theorem 4.2 *The expected missing event detection time $\mathbb{E}[T_D]$ is given by the following equation:*

$$\mathbb{E}[T_D] = \frac{|\mathbb{E}|_{t_r} x}{\ln 2} + t_t \sum_{N^* = |\mathbb{E}| - M}^{|\mathbb{E}| - M + |\mathbb{U}|} \frac{1 - (1 - q)^f - f q (1 - q)^f}{q}$$

$$\binom{|\mathbb{U}|}{N^* - |\mathbb{E}| + M}\left(\frac{1}{2^x}\right)^{N^* - |\mathbb{E}| + M}\left(1 - \frac{1}{2^x}\right)^{|\mathbb{U}| - N^* + |\mathbb{E}| - M}. \qquad (4.26)$$

Proof Recall (4.16), it holds that there are $f = \frac{-N^* y^*}{\ln(1 - (1 - \alpha)^{\frac{1}{M y^*}})}$ slots in Phase 2. We next calculate the number of slots before detecting the first missing tag. It is easy to check that the event that in slot z the reader detects the first missing tag happens if no missing tags is detected in the first $z - 1$ slots while at least one missing tag is detected in slot z. Let Z denote the random variable of z, we have

$$P\{Z = z\} = (1 - q)^{z-1} * q, \tag{4.27}$$

which is geometrically distributed.

We can then compute the expectation of Z, conditioned by N^*, as follows:

$$E[Z|N^*] = \sum_{z=1}^{f} z \cdot P\{Z = z\}$$

$$= \frac{1 - (1 - q)^f - fq(1 - q)^f}{q}. \tag{4.28}$$

Moreover, it follows from the analysis of Phase 1 that the probability that an unexpected tag is still active after Phase 1 is $\prod_{j=1}^{J} P_{1,j}$. On the other hand, since $\mathbb{U}_r$ represents the ID set of active unknown tags after Phase 1, recall (4.5) and $\sum_{j=1}^{J} k_j^* = x$, we can compute the probability of having u active unexpected tags after Phase 1 as follows:

$$P\{|\mathbb{U}_r| = u\} = \binom{|\mathbb{U}|}{u} \left(\prod_{j=1}^{J} P_{1,j}\right)^u \left(1 - \prod_{j=1}^{J} P_{1,j}\right)^{|\mathbb{U}|-u}$$

$$= \binom{|\mathbb{U}|}{u} \left(\frac{1}{2^x}\right)^u \left(1 - \frac{1}{2^x}\right)^{|\mathbb{U}|-u}. \tag{4.29}$$

It can be noted that $|\mathbb{U}_r|$ follows the binomial distribution. Recall the relationship between N^* and $|\mathbb{U}_r|$ in (4.7), it holds that

$$E[Z] = \sum_{N^*=|\mathbb{E}|-M}^{|\mathbb{E}|-M+|\mathbb{U}|} E[Z|N^*] \binom{|\mathbb{U}|}{N^* - |\mathbb{E}| + M}.$$

$$\left(\frac{1}{2^x}\right)^{N^*-|\mathbb{E}|+M} \cdot \left(1 - \frac{1}{2^x}\right)^{|\mathbb{U}|-N^*+|\mathbb{E}|-M} \tag{4.30}$$

Therefore, $E[T_D]$ can be derived as

$$E[T_D] = T_1 + E[Z] \cdot t_t = \frac{|\mathbb{E}|t_r x}{\ln 2} + E[Z] \cdot t_t. \tag{4.31}$$

Injecting $E[Z]$ into $E[T_D]$ completes the proof.   □

After deriving $E[T_D]$ as a function of x, we seek the optimum, denoted by x_e^*, which minimizes $E[T_D]$. To this end, we first establish an upper-bound of x_e^* in the following lemma.

Lemma 4.3 *It holds that* $x_e^* \leq \frac{2t_t \ln 2}{t_r |\mathbb{E}| q_{min}}$.

Proof We write $E[T_D]$ as a function of x. Specifically, let $E[T_D] = g(x)$. To prove the lemma, we show that for any $x > 2x_0$ it holds that $g(x) \geq g(x_0)$ where $x_0 \triangleq \frac{t_t \ln 2}{t_r |\mathbb{E}| q_{min}}$.

To this end, we first derive the bounds of $g(x)$. Recall (4.27),(4.28), (4.30) and (4.31), we have

$$g(x) > \frac{|\mathbb{E}| t_r x}{\ln 2},$$

$$g(x) \leq \frac{|\mathbb{E}| t_r x}{\ln 2} + \frac{t_t}{q_{min}}.$$

For any $x > 2x_0$, we then have

$$g(x) > \frac{|\mathbb{E}| t_r x}{\ln 2} > \frac{2|\mathbb{E}| t_r x_0}{\ln 2} = \frac{|\mathbb{E}| t_r x_0}{\ln 2} + \frac{t_t}{q_{min}} \geq g(x_0)$$

The lemma is thus proved. □

Lemma 4.3 shows that x_e^* falls into the range $[0, 2x_0]$. We can thus search $[0, 2x_0]$ to find x_e^* that minimises $E[T_D]$ and then set J and k_j^* such that $\sum_{j=1}^{J} k_j^* = x_e^*$.

4.5.5 BMTD Parameter Setting: Summary

We conclude this section by streamlining the procedure of the parameter setting in the BMTD:

1. *Set parameters in Phase 2:* given $|\mathbb{E}|$, M, α and $|\mathbb{U}|$, compute f_w^* and y^* by (4.17) and (4.18), respectively, and set R_w and W such that $R_w W = y^*$;
2. *Set parameters in Phase 1:* compute x^* by Theorem 4.1 if the objective is to minimise the worst-case execution time; compute x_e^* if the objective is to minimise the expected detection time; then the set of k_j^* and J is given such that $\sum_{j=1}^{J} k_j^* = x^*$ or $\sum_{j=1}^{J} k_j^* = x_e^*$.

Following the above two steps, we can obtain all parameters in the BMTD. Note that R_w and W can be picked arbitrarily as long as $R_w W = y^*$ is satisfied, if we set $R_w = 1$ and $W = y^*$, then the Phase 2 of BMTD is reduced to RUN [5], RUN is thus a special case of our proposed BMTD.

4.6 Cardinality Estimation

In order to execute the BMTD, the reader needs to estimate the number of unexpected tags $|\mathbb{U}|$. In our work, we use the SRC estimator which is designed in [16] and is the current state-of-the-art solution. Denote by $\overline{|\mathbb{E}| - m + |\mathbb{U}|}$ the estimated total number of tags in the system, then the cardinality $|\mathbb{U}|$ can be approximated as $\overline{|\mathbb{U}|} = \overline{|\mathbb{E}| - m + |\mathbb{U}|} - |\mathbb{E}|$ if $m << |\mathbb{E}|, |\mathbb{U}|$. Because the number of bits that set to one in Bloom filter is concentrated tightly around the mean [22] and [23], once the estimation $\overline{|\mathbb{U}|}$ is obtained, we can calculate the expectation of N^* according to (4.9) with $m = M$ and use it as the estimator of N^*.

The SRC estimator consists of two phases: rough estimation and accurate estimation. It is proven in [16] that SRC can obtain a rough estimation $\hat{n}$ which at least equals to $0.5(|\mathbb{E}| - m + |\mathbb{U}|)$ after its first phase. In the second phase, SRC can achieve that the relative estimation error is not greater than ϵ which is referred to as confidence range with the settings as follows: the frame size $L_{est} = \frac{65}{(1-0.04\epsilon)^2}$ and the persistence probability $p_{pe} = \min\{1, 1.6L_{est}/\hat{n}\}$.

We then analyse the overhead introduced to estimate the cardinality of $\mathbb{U}$. As proven in [16], the overhead of SRC estimator is at most $O(\frac{1}{\epsilon^2} + \log\log(|\mathbb{U}| + |\mathbb{E}|))$, which is moderate for large-scale RFID systems with large $|\mathbb{U}|$ and $|\mathbb{E}|$.

4.6.1 Fast Detection of Missing Event

In our estimation approach, we require that $m \ll |\mathbb{E}|, |\mathbb{U}|$. In case where m is close to $|\mathbb{E}|, |\mathbb{U}|$, the estimation may not be accurate. Luckily, in this case, we can quickly detect a missing event in the cardinality estimation phase due to large m.

Specifically, we analyze the SRC estimator's capability of detecting missing event under large m by comparing the pre-computed slots with those selected by the present tags. Recall the proof of Lemma 4.2, we can derive the detection probability in any given slot, defined as q_{pre}, as

$$q_{pre} = \left(1 - \left(1 - \frac{p_{pe}}{L_{est}}\right)^m\right) * \left(1 - \frac{p_{pe}}{L_{est}}\right)^{(\mathbb{U}+\mathbb{E}-m)}. \tag{4.32}$$

Since the detections in different slots are independent of each other, the probability of detecting at least one missing tag event by the SRC estimator can be calculated as $1 - (1 - q_{pre})^{L_{est}}$ which is a increasing function of m.

Figure 4.3 illustrates the detection probability of SRC with the various number of missing tags under different unexpected tag population sizes. To obtain the figure, we set $|\mathbb{E}| = 10^3$ and $\epsilon = 0.1$. It is observed that in the cases that $|\mathbb{U}| = 0.5 * 10^4$, $1 * 10^4, 2 * 10^4$, SRC is able to detect at least a missing tag event with probability one when m is not less than 100, 200, 600, which means that a missing event is detected by SRC and the reader does not need to invoke the BMTD. In the other side, in the cases that m is less than 100, 200, 600, it holds that $|\frac{\overline{\mathbb{U}}}{\mathbb{U}} - 1| \leq 0.138, 0.132, 0.128$, respectively. With reference to the conclusion drawn from the Fig. 4.4, the BMTD can tolerate these levels of estimation error.

Fig. 4.3 q_{pre} vs. m

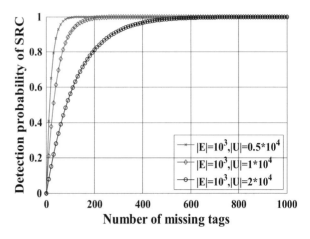

Fig. 4.4 $\left|\frac{\overline{E[T_D]}}{E[T_D]} - 1\right|$ vs. $\frac{\overline{|U|}}{|U|}$

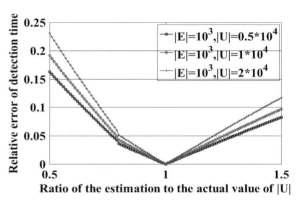

4.6.2 Sensibility to Estimation Error

The estimation algorithm we use inevitably introduces error on $|\mathbb{U}|$, which may have a negative impact on the performance of the BMTD. In order to investigate this impact, we next illustrate the sensitivity of the detection time to the estimation error.

Figure 4.4 shows the theoretically calculated expected detection time from (4.26) under different unexpected tag population sizes and various levels of estimation error for $M = 1$. All results here are normalized with respect to the expected detection time without estimation error, which can be represented as $\left|\frac{\overline{E[T_D]}}{E[T_D]} - 1\right|$. As shown in the figure, the relative error of detection time increases with the estimation error in all range of unexpected tag population. But it is worth noticing that the relative error of detection time only increases by 5% at most when $\left|\frac{\overline{|U|}}{|U|} - 1\right| \leq 0.2$, which is nearly same with that without estimation error. Note that the slope before $\frac{\overline{|U|}}{|U|} = 1$ changes because the calculated x_e changes from 2 to 3 which is optimal.

4.6.3 Enforcing Detection Reliability

Estimation error also has impact on the reliability of the BMTD as P_{sys} is calculated base on the estimated cardinality.

To enforce the detection reliability, we introduce more rounds to execute additional Bloom filters. The scheme works as follows: After receiving the Bloom filtering vector constructed by the active tags in the set $\mathbb{E}_r \cup \mathbb{U}_r$ in each round of Phase 2, the reader first counts the actual number of '1' bits in the filtering vector, defined as s_1 and uses it to compute the actual false positive probability, denoted by $\hat{P}_{2,w}$, as follows:

$$\hat{P}_{2,w} = \frac{s_1}{f_w^*}, \tag{4.33}$$

because an arbitrary unexpected tag maps to a '1' bit with a probability of s_1 out of f_w^*.

Following (4.13), we have the observed protocol reliability, denoted by $\hat{P}_{sys}$, as follows:

$$\hat{P}_{sys} = 1 - \hat{P}_{2,w}^{MW}. \tag{4.34}$$

If $\hat{P}_{sys} < \alpha$, the reader adds one more round in Phase 2 to further detect the missing tag event until $\hat{P}_{sys} \geq \alpha$.

4.6.4 Discussion on Multi-Reader Case

In large-scale RFID systems deployed in a large area, multiple readers are thus deployed to ensure the full coverage for a larger number of tags in the interrogation region. In such scenarios, we leverage the approach proposed in [24] and employed in [5]. The main idea is that a back-end server is used to synchronize all readers such that the RFID system with multiple readers operates as the single-reader case.

Specially, the back-end server calculates all the parameters involved in BMTD and constructs Bloom filter and sends them to all readers such that they broadcast the same parameters and Bloom filter to the tags. Furthermore, each reader sends its individual Bloom filtering vector back to the back-end server. When the back-end server receives all Bloom filtering vectors, it applies logical OR operator on all received Bloom filtering vectors, which eliminates the impact of the duplicate readings of tags in the overlapped interrogation region. Consequently, a virtual Bloom filter is constructed by the back-end server.

4.7 Performance Evaluation

The problem addressed in this chapter is to detect the missing expected tags in the presence of a large number of unexpected tags in a time-efficient and reliable way. In this section, we evaluate the performance of the proposed BMTD. It has been shown in [5] that existing missing detection protocols cannot achieve the required reliability when there are unexpected tags in the RFID systems except the latest RUN [5]. We thus compare our proposed BMTD to RUN in terms of the actual reliability and the detection time. Note that the detection time can be interpreted as the time taken to either detect the fist missing tag event if a missing tag is found or complete the execution if no missing tag is found.

The simulation parameters are set with reference to [4] and [5]. Specifically, since both transmission rates from the tags to the reader and the reader to the tags depend on physical implementation and interrogation environment, we make the same assumption as in [4] that $t_r = t_t$. Moreover, because RUN is the baseline protocol, we use the similar simulation scenarios and the same performance metrics as in [5] where the time needed to detect a missing tag event is shown in terms of the number of slots. To that end, we, without loss of generality, assume $t_r = t_t = 1$ in (4.26) in the simulation. Besides, we compute the optimal parameter values for RUN by following its specifications.

In the simulation, we use SRC [16] armed with missing tag detection function in this chapter to estimate the unexpected tag population size with the confidence rang $\epsilon = 0.1$. And all presented results are obtained by taking the average value of 100 independent trials under the same simulation setting.

We start by evaluating the performance of the BMTD by optimizing the worst-case execution time and the expected detection time.

4.7.1 Comparison Between Two Strategies of BMTD

In this subsection, we compare the performance of two strategies of the BMTD which are abbreviated to Worst-M and Expected-M here, respectively. We set $|\mathbb{E}| = 1000$, $m = 100$, $\alpha = 0.9$, $|\mathbb{U}| = 10,000 : 5000 : 30,000$, $M = 1$ and 50.

Table 4.2 lists the results where the first and second elements in the two-tuple $(\cdot, \cdot)$ denote the actual reliability and detection time, respectively. It can be seen that Expected-M costs less time than Worst-M to achieve the same reliability which is greater than the system requirement on the detection reliability, especially when M is small. Specifically, compared with Worst-1, Expected-1 reduces the detection time by up to 51.92% when $|\mathbb{U}| = 10,000$. This is because $x^* = 5$ is too large for Phase 1 by optimizing the worst-case execution time, which wastes time. In contrast, minimizing the expected detection time relieves the influence of unexpected tag population size on the time of Phase 2 and thus outputs a smaller $x_e^* = 2$. In the rest of our simulation, we configure the parameters of the BMTD to minimise the expected detection time.

Table 4.2 Actual reliability and detection time of BMTD

Strategy	Number of unexpected tags				
	10,000	15,000	20,000	25,000	30,000
Worst-1	(1, 4108)	(1, 4441)	(1, 5013)	(1, 5453)	(1, 5510)
Expected-1	(1, 1975)	(1, 3187)	(1, 3569)	(1, 3828)	(1, 4191)
Worst-50	(1, 1357)	(1, 1841)	(1, 2753)	(1, 2762)	(1, 2995)
Expected-50	(1, 1353)	(1, 1618)	(1, 2272)	(1, 2472)	(1, 2815)

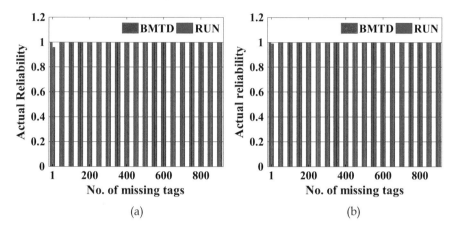

Fig. 4.5 Actual reliability vs. number of missing tags. (**a**) $\alpha = 0.9$, (**b**) $\alpha = 0.99$

4.7.2 Comparison between BMTD and RUN

4.7.2.1 Comparison Under Different Number of Missing Tags

In this subsection, we evaluate the performance of BMTD under different number of missing tags, which stands for the effectiveness and efficiency of BMTD. To that end, we set $|\mathbb{E}| = 1000$, $|\mathbb{U}| = 30,000$, $m = 1 : 50 : 901$, $\alpha = 0.9$ and 0.99. Moreover, we set the threshold to $M = 1$.

Actual Reliability BMTD achieves the required reliability for any missing tag population size when there are a large number of unexpected tags in the RFID systems. Figure 4.5a, b illustrate the actual reliability of BMTD and RUN for $\alpha = 0.9$ and 0.99, respectively. It can be observed that both BMTD and RUN achieve the reliability more than that required by the system.

Detection Time BMTD is more time-efficient in comparison to RUN. Figure 4.6a, b show the detection time for $\alpha = 0.9$ and 0.99, respectively. For clearness, we further highlight the caves from $m = 51$ to 901. As shown in the figures, the detection time of BMTD is far shorter than that of RUN and decreases with the number of missing tags significantly. This is unsurprising. BMTD is able to deactivate major unexpected tags, which greatly reduces the number of active tags in the population, such that the presence of more missing tags makes the detection

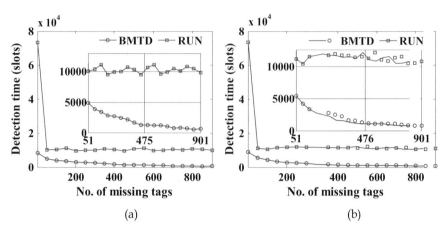

Fig. 4.6 Detection time vs. number of missing tags. (**a**) $\alpha = 0.9$, (**b**) $\alpha = 0.99$

much easier. In contrast, RUN does not take into account the impact of unexpected tag population size, leading to longer detection delay in the presence of large number of unexpected tags.

4.7.2.2 Comparison Under Different Number of Unexpected Tags

In this subsection, we evaluate the performance of BMTD under different number of unexpected tags, which represents the generality of BMTD. To that end, we set $|\mathbb{E}| = 1000$, $m = 50$, $M = 1$, $\alpha = 0.9$ and 0.99. Moreover, we select such $|\mathbb{U}| = 1000, 5000 : 5000 : 30,000$ that various values of $\frac{|\mathbb{U}|}{|\mathbb{E}|}$ are covered in the simulation.

Actual Reliability BMTD achieves the reliability greater than the required reliability for different cardinalities of unexpected tag set. Figure 4.7a, b depict the actual reliability of BMTD and RUN for $\alpha = 0.9$ and 0.99, respectively. It can be observed that the actual reliability achieved by both BMTD and RUN is equal to one.

Detection Time The BMTD outperforms the RUN considerably in terms of detection time even in the scenario with the small number of unexpected tag. Figure 4.8a, b show the detection time for $\alpha = 0.9$ and 0.99, respectively. As shown in the figures, BTMD is able to save time especially when more unexpected tags are present in the population. Moreover, the increase in detection time of BTMD is more slow than that of RUN. This is due to the ability of BTMD that it can detect the missing tag event when estimating the $|\mathbb{U}|$ and determine whether to execute the unexpected tag deactivation phase following Lemma 4.3, which is exactly ignored in RUN.

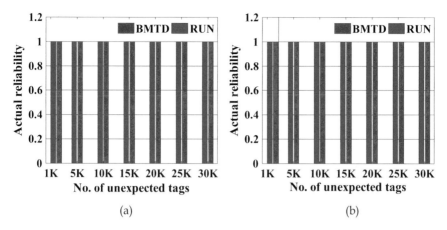

Fig. 4.7   Actual reliability vs. number of unexpected tags. (**a**) $\alpha = 0.9$, (**b**) $\alpha = 0.99$

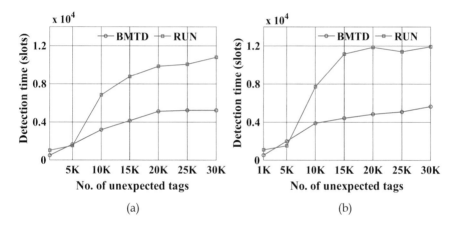

Fig. 4.8   Detection time vs. number of unexpected tags. (**a**) $\alpha = 0.9$, (**b**) $\alpha = 0.99$

4.7.2.3   Comparison Under Different Values of Threshold

In this subsection, we evaluate the performance of BMTD under different thresholds, which represents the tolerability of BMTD. To that end, we set $|\mathbb{E}| = 1000$, $|\mathbb{U}| = 30{,}000$, $m = 100$, $\alpha = 0.9$ and 0.99. Moreover, we choose such $M = 50$: $50 : 300$ that the threshold can be greater or smaller than or equal to the number of missing tags in the simulation.

Actual Reliability   BMTD achieves better reliability than the required reliability when $m \geq M$. As shown in Fig. 4.9a, b, BMTD fails to achieve the required reliability only when $m < M$, which does not have negative impact because the objective of the missing tag detection protocol is to detect the missing tags only if the number of missing tags exceeds the threshold M.

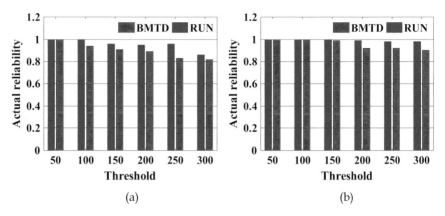

Fig. 4.9   Actual reliability vs. threshold. (**a**) $\alpha = 0.9$, (**b**) $\alpha = 0.99$

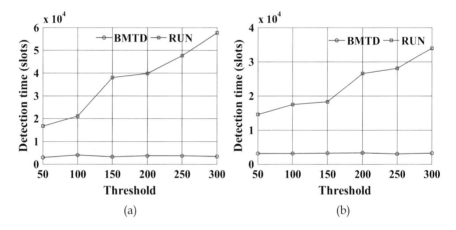

Fig. 4.10   Detection time vs. threshold. (**a**) $\alpha = 0.9$, (**b**) $\alpha = 0.99$

Detection Time   BMTD can tolerate the deviation from the threshold in terms of the detection time even when $m < M$. Figure 4.10a, b show the detection time for $\alpha = 0.9$ and 0.99, respectively. It can be seen from the figures that the detection time of BMTD almost does not vary with the deviation. The detection time of RUN, by contrast, increases substantially as the deviation increase when $m < M$. This is because RUN terminates only when it runs optimal number of frames since the first frame when the estimated value of $|\mathbb{U}|$ does not vary by 0.1% in consecutive 10 frames if it does not detect any missing tag in any frame, while BMTD stops once the observed reliability $\hat{P}_{sys}$ exceeds α.

4.8 Conclusion

This chapter has investigated an important problem of detecting missing tags in the presence of a large number of unexpected tags in large-scale RFID systems. Specifically, we aim at detecting a missing tag event in a reliable and time-efficient way. This chapter has presented a two-phase Bloom filter-based missing tag detection protocol (BMTD). In the first phase, we employed Bloom filter to screen out and then deactivate the unexpected tags in order to reduce their interference to the detection. In the second phase, we further used Bloom filter to test the membership of the expected tags to detect missing tags. We also showed how to configure the protocol parameters so as to optimize the detection time with the required reliability. Furthermore, we conducted extensive simulation experiments to evaluate the performance of the proposed protocol and the results demonstrate the effectiveness and efficiency of the propose protocol in comparison with the state-of-the-art solution.

References

1. National Retail Federation, National retail security survey. [Online] (2015), https://nrf.com/resources/retail-library/national-retail-security-survey-2015
2. C.C. Tan, B. Sheng, Q. Li, How to monitor for missing RFID tags, in *IEEE ICDCS* (IEEE, Piscataway, 2008), pp. 295–302
3. W. Luo, S. Chen, T. Li, Y. Qiao, Probabilistic missing-tag detection and energy-time tradeoff in large-scale RFID systems, in *ACM MobiHoc* (ACM, New York, 2012), pp. 95–104
4. W. Luo, S. Chen, Y. Qiao, T. Li, Missing-tag detection and energy–time tradeoff in large-scale RFID systems with unreliable channels. IEEE/ACM Trans. Netw. **22**(4), 1079–1091 (2014)
5. M. Shahzad, A.X. Liu, Expecting the unexpected: Fast and reliable detection of missing RFID tags in the wild, in *IEEE INFOCOM* (2015), pp. 1939–1947
6. T. Li, S. Chen, Y. Ling, Identifying the missing tags in a large RFID system, in *ACM MobiHoc* (ACM, New York, 2010), pp. 1–10
7. R. Zhang, Y. Liu, Y. Zhang, J. Sun, Fast identification of the missing tags in a large RFID system, in *IEEE SECON* (IEEE, Piscataway, 2011), pp. 278–286
8. X. Liu, K. Li, G. Min, Y. Shen, A.X. Liu, W. Qu, Completely pinpointing the missing RFID tags in a time-efficient way. IEEE Trans. Comput. **64**(1), 87–96 (2015)
9. J. Myung, W. Lee, Adaptive splitting protocols for RFID tag collision arbitration, in *ACM MobiHoc* (ACM, New York, 2006), pp. 202–213
10. V. Namboodiri, L. Gao, Energy-aware tag anticollision protocols for rfid systems. IEEE Trans. Mob. Comput. **9**(1), 44–59 (2010)
11. T.F. La Porta, G. Maselli, C. Petrioli, Anticollision protocols for single-reader RFID systems: temporal analysis and optimization. IEEE Trans. Mob. Comput. **10**(2), 267–279 (2011)
12. M. Shahzad, A.X. Liu, Probabilistic optimal tree hopping for RFID identification, in *ACM SIGMETRICS*, vol. 41 (ACM, New York, 2013), pp. 293–304
13. C. Qian, H. Ngan, Y. Liu, L.M. Ni, Cardinality estimation for large-scale RFID systems. IEEE Trans. Parallel Distrib. Syst. **22**(9), 1441–1454 (2011)
14. M. Shahzad, A.X. Liu, Every bit counts: fast and scalable RFID estimation, in *ACM Mobicom* (2012), pp. 365–376

15. Y. Zheng, M. Li, Zoe: fast cardinality estimation for large-scale RFID systems, in *IEEE INFOCOM* (IEEE, Piscataway, 2013), pp. 908–916
16. B. Chen, Z. Zhou, H. Yu, Understanding RFID counting protocols, in *ACM MobiHoc* (2013), pp. 291–302
17. B.H. Bloom, Space/time trade-offs in hash coding with allowable errors. Commun. ACM **13**(7), 422–426 (1970)
18. O. Rottenstreich, Y. Kanizo, I. Keslassy, The variable-increment counting bloom filter. IEEE/ACM Trans. Netw. **22**(4), 1092–1105 (2014)
19. F. Hao, M. Kodialam, T. Lakshman, H. Song, Fast dynamic multiple-set membership testing using combinatorial bloom filters. IEEE/ACM Trans. Netw. **20**(1), 295–304 (2012)
20. H. Han, B. Sheng, C.C. Tan, Q. Li, W. Mao, S. Lu, Counting RFID tags efficiently and anonymously, in *IEEE INFOCOM* (IEEE, Piscataway, 2010), pp. 1–9
21. EPCglobal Inc., Radio-frequency identity protocols class-1 generation-2 UHF RFID protocol for communications at 860 mhz - 960 mhz version 1.0.9. [Online]
22. M. Mitzenmacher, E. Upfal, *Probability and Computing: Randomized Algorithms and Probabilistic Analysis* (Cambridge University Press, Cambridge, 2005)
23. F. Hao, M. Kodialam, T. Lakshman, Building high accuracy bloom filters using partitioned hashing, in *ACM SIGMETRICS* (ACM, New York, 2007), pp. 277–288
24. M. Kodialam, T. Nandagopal, W.C. Lau, Anonymous tracking using RFID tags, in *IEEE INFOCOM* (IEEE, Piscataway, 2007), pp. 1217–1225

Chapter 5
On Missing Tag Detection in Multiple-Group Multiple-Region RFID Systems

Chapter Roadmap The rest of this chapter is organised as follows. Section 5.1 explains the motivation of studying missing tag detection in Multi-group multi-region systems. Section 5.2 formulates the missing tag detection problem in RFID systems with the presence of unexpected tags. Section 5.3 details the baseline missing tag detection protocol. Section 5.4 introduces the adaptive Bloom-filter based protocol. Section 5.5 proposes a group-wise protocol to further reduce execution time. Section 5.7 shows the experimental results. Section 5.8 gives the summary.

5.1 Introduction

We investigate a different problem in this chapter motivated by the increasing application of mobile reader [1, 2] and the following practical settings.

- *Multiple groups of tags.* Tags are usually attached to objects belonging to different groups: e.g., different brands of the goods with the high-end brands order-of-magnitude more valuable than their low-end peers. Therefore, the missing tag events are characterized by asymmetrical threshold and reliability requirement across groups.
- *Multiple interrogation regions.* Tags may be unevenly located in multiple interrogation regions: e.g., tags may be located in several rooms or different corners or regions of a large warehouse. Hence, a reader may need to move several times to cover all monitored tags and complete the missing tag detection process.

The problem we consider is to devise missing tag detection protocol with minimum execution time while guaranteeing the detection reliability requirement for each group of tags in multiple-region scenario. In the considered multiple-group

© Springer International Publishing AG, part of Springer Nature 2019   105
J. Yu, L. Chen, *Tag Counting and Monitoring in Large-Scale RFID Systems*,
https://doi.org/10.1007/978-3-319-91992-8_5

multiple-region scenario, all existing missing tag detection protocols cannot work effectively due to the following two reasons. First, existing approaches require the full coverage of tags when executing the detection algorithms, which clearly does not hold in the considered multiple-region scenario. Secondly, existing work does not take into account the heterogeneity among groups and thus either cannot meet the individual reliability requirement, or suffers extremely long detection delay.

To solve this challenging problem, we deliver a comprehensive analysis on the missing tag detection problem in the above multiple-group multiple-region environment and investigate how to devise optimum missing tag detection algorithms. Note that when there are only one group and all tags are with one interrogation region, our problem degenerates to the classical missing tag detection problem studied in the literature.

To design missing tag detection algorithms in the multiple-region multiple-group case, we leverage a powerful technique called *Bloom filter* which is a space-efficient probabilistic data structure for representing a set and supporting set membership queries [3] to detect a missing event. Specifically, we develop a suite of three missing tag detection protocols, each decreasing the execution time compared to its predecessor by incorporating an improved version of the Bloom filter design and parameter tuning. By sequentially analysing the developed protocols, we gradually iron out an optimum detection protocol that works in practice.

5.2   System Model and Problem Formulation

5.2.1   System Model

We consider a grouped RFID system composed of a mobile reader and G groups of tags distributed in R ($R \geq 1$) interrogation regions (e.g., R rooms), concisely referred to as regions. In case where a tag may be physically located in two regions, i.e., regions may overlap one with another, the tag only responses to reader queries regarding to the first region when it is interrogated. In this sense, we can treat the regions as non-overlapping ones.

We use $\mathbb{E}$ to denote the set of the tags which are expected to be present and we denote its cardinality (i.e., the number of expected tags) by $|\mathbb{E}|$. The reader knows the IDs of all tags in $\mathbb{E}$ but does not know the set of tags in each region. For presentation conciseness, we set the ID of group g ($1 \leq g \leq G$) to its index g. We assume every tag knows its group ID through a grouping protocol, e.g. [4]. We also assume the reader knows the approximate number of tags of each group g actually present in each region r ($1 \leq r \leq R$), denoted by n_{gr}. The estimation of n_{gr} can be achieved by the reader by deactivating all tags not belonging to group g (using the ID of group g) and then using any state-of-the-art tag population estimation algorithm.

Table 5.1 Main notations

Symbols	Descriptions
G	Number of groups
g	Group index and group ID
R	Number of interrogation regions
r	Region index
$\mathbb{E}$	Set of target tags that need to be monitored
n_{gr}	Number of tags of group g in region r
m_g	Number of missing tags in group g
M_g	Threshold of group g
P_{dg}	Probability of detecting a missing event of group g
α_g	System requirement on the detection reliability for group g
f	Length of Bloom filter in B-detect
k	Number of hash functions in B-detect
s	Hash function seed
P_{fp}	False positive rate of Bloom filter in B-detect
T_B	Execution time of B-detect
f_r	Bloom filter vector size in region r in AB-detect
k_g	Number of hash functions for group g in AB/GAB-detect
$P_{fp,g}$	False positive rate of Bloom filter for g in AB/GAB-detect
T_{AB}	Execution time of AB-detect
f_{gr}	Bloom filter vector length for group g in r in GAB-detect
T_{GAB}	Execution time of GAB-detect

Table 5.1 summaries main notations used in this chapter.

5.2.2   Problem Formulation

We are interested in detecting missing tag event for each group g. Let m_g denote the number of missing tags in group g which is of course not known by the reader. Let M_g denote the threshold of group g. A missing event of group g denotes the event where there are at least M_g tags of group g missing in the system. Let P_{dg} denote the probability that the reader can detect a missing event of group g, we formulate the optimum missing tag detection problem as follows.

Definition 5.1 (Optimum Missing Tag Detection Problem) The optimum missing tag detection problem is to devise an algorithm of minimum execution time which can detect a missing event for each group g with probability $P_{dg} \geq \alpha_g$ if $m_g \geq M_g$, where α_g is the requirement on the detection reliability for group g. When there is only one group in the system, the problem degenerates to the classical missing event detection problem.

5.2.3 Design Rational

To design missing tag detection algorithms in the multiple-region multiple-group case, we leverage a powerful technique called *Bloom filter* which is a space-efficient probabilistic data structure for representing a set and supporting set membership queries [3] to detect a missing event. In our design, we explore the following three natural ideas, each corresponding to a proposed missing tag detection protocol detailed in the next three sections.

Baseline Approach To enable missing tag detection in the multiple-region multiple-group case, we let the reader use the same Bloom filter parameters in each region for each group of tags and construct the Bloom filter based on the responses from the tags to perform missing event detection. This approach, termed as *B-detect*, is a direct application of Bloom filter to solve our problem.

Adaptive Approach In the baseline approach B-detect, the reader uses the same parameters in each region, which may not be optimum in the case when tags are not evenly distributed across regions. Motivated by this observation, we develop an adaptive approach, named *AB-detect*, which enables the reader to use different parameters based on the number of tags in the interrogation region the reader queries. Specifically, for each region r, the reader executes one query, to which tags of all the groups in the region respond. The reader constructs a Bloom filter B_r for each region containing the response and aggregates B_r ($1 \le r \le R$) to form a virtual Bloom filter B^{AB}, based on which it detects missing event for each group.

Group-Wise Approach We go further by developing a group-wise approach, referred to as *GAB-detect*. In GAB-detect, the reader executes G group-wise queries for each region r. Only tags of group g ($1 \le g \le G$) in the interrogation region respond to the g-th query. The reader then constructs a Bloom filter B_{gr}^{GAB} for each group g and aggregates B_{gr}^{GAB} ($1 \le r \le R$) to form a virtual Bloom filter B_{g*}^{GAB} using the technique in AB-detect, based on which it detects missing event for group g.

   By sequentially analysing the above three approaches and mathematically comparing their performance in terms of execution time, we gradually iron out an optimum detection protocol that works in practice.

5.3 The Baseline Approach

In the B-detect design to enable missing tag detection in the multiple-region case, we let the reader use the same parameters in each region and construct the Bloom filter based on the responses from the tags to perform missing event detection. Specifically, B-detect consists of two phases, detailed as below.

5.3.1   Protocol Description

Phase 1: Query and Feedback Collection   The reader performs a query in each region r with the same parameter setting (f, k, s), where f is the length of the Bloom filter vector, k is the number of independent hash functions used to construct the Bloom filter vector, and s is the seed of the hash functions which is identical for all groups and regions. How their values are chosen is analysed in Sect. 5.3.2 on parameter optimisation. Upon receiving the request, each tag in region r, regardless of the group to which it belongs, selects k slots $(h_v(ID) \mod f) \, (1 \leq v \leq k)$ in the frame of f slots and replies in these slots. The reader then constructs a Bloom filter vector B_r with the responses from the tags in each region r as follows. Note there are two types of slots: empty slots and nonempty slots. According to the responses from tags, if slot i $(1 \leq i \leq f)$ is empty, the reader sets $B_r(i) = 0$, otherwise it sets $B_r(i) = 1$.

Phase 2: Virtual Bloom Filter Construction and Missing Event Detection   After interrogating all R regions, the reader combines the Bloom filter vectors B_r $(1 \leq r \leq R)$ to a virtual Bloom filter B by ORing each bit of them, i.e., $B(i) = B_1(i) \vee \cdots \vee B_R(i)$. The reader then performs membership test. For each tag in $\mathbb{E}$, the reader maps its ID into k bits at positions $(h_v(ID) \mod f) \, (1 \leq v \leq k)$. If all the corresponding bits in B are 1, then the tag is regarded as present. Otherwise, the tag is considered to be missing. The reader reports a missing event in group g if the number of missing tags is at least M_g and no missing event otherwise.

5.3.2   Performance Optimisation and Parameter Tuning

The execution time of B-detect, defined as T_B in number of slots, can be written as

$$T_B = R(t_1 + f\delta) \simeq Rf\delta, \tag{5.1}$$

where t_1 denotes the time for the reader to broadcast the query parameters and δ denotes the slot duration which we normalise to 1 for notation conciseness. In a large RFID system, it holds that $f \gg t_1$, so we ignore t_1. In this subsection, we derive the optimum value of f that minimizes T_B.

It is well-known that there is no false negative in the Bloom filter membership test and the false positive rate P_{fp} for an arbitrary group g can be calculated as follows [3]:

$$P_{fp} = \left[1 - \left(1 - \frac{1}{f} \right)^{(|\mathbb{E}| - m)k} \right]^k \approx (1 - e^{-(|\mathbb{E}| - m)k/f})^k, \tag{5.2}$$

where $m = \sum_{g=1}^{G} m_g$ denotes the total number of missing tags in all groups.

By rearranging (5.2), we can express the Bloom filter size as

$$f = \frac{-(|\mathbb{E}| - m)k}{\ln(1 - P_{fp}^{\frac{1}{k}})}. \tag{5.3}$$

The following theorem derives the optimal values of f and k in the sense of minimising the execution time.

Theorem 5.1 *The optimum size of the Bloom filter and the optimum number of hash functions in B-detect, denoted by f^* and k^* respectively, that minimize the execution time while satisfying the detection reliability requirement for each group g regardless of m_g, are as follows:*

$$f^* = (|\mathbb{E}| - M) \cdot \frac{k^*}{-\ln(1 - X_{g^*}^{\frac{1}{k^*}})}, \tag{5.4}$$

$$k^* = \frac{\ln\left(1 - \alpha_{g^*}^{\frac{1}{M_{g^*}}}\right)}{\ln \frac{1}{2}}, \tag{5.5}$$

where $M = \sum_{g=1}^{G} M_g$, $X_g \triangleq 1 - \alpha_g^{\frac{1}{M_g}}$, and $g^* = \arg\min_g X_g$.

Proof Recall the definition of a missing event in group g that at least M_g tags are missing, the probability that a missing event can be detected in group g by the reader, defined as P_{dg}, can be computed as

$$P_{dg} = \sum_{i=M_g}^{m_g} \binom{m_g}{i}(1 - P_{fp})^i P_{fp}^{m_g - i}, \tag{5.6}$$

and P_{dg} has the following property for any $m_g \geq M_g$:

$$P_{dg} = (1 - P_{fp})^{M_g} \sum_{i=M_g}^{m_g} \binom{m_g}{i}(1 - P_{fp})^{i - M_g} P_{fp}^{m_g - i}$$

$$= (1 - P_{fp})^{M_g} \sum_{j=0}^{m_g - M_g} \binom{m_g}{j + M_g}(1 - P_{fp})^j P_{fp}^{m_g - M_g - j}$$

$$\geq (1 - P_{fp})^{M_g} \sum_{j=0}^{m_g - M_g} \binom{m_g - M_g}{j}(1 - P_{fp})^j P_{fp}^{m_g - M_g - j}$$

$$\geq (1 - P_{fp})^{M_g}, \tag{5.7}$$

where the first inequality holds due to the inequality below

$$\frac{\binom{m_g}{j+M_g}}{\binom{m_g-M_g}{j}} = \prod_{i=0}^{M_g-1} \frac{m_g - i}{M_g + j - i} \geq 1, \ \forall j \in [0, m_g - M_g],$$

where the equality holds when $m_g = M_g$.

Hence, to ensure the system requirement $P_{dg} \geq \alpha_g$ regardless of m_g, we must ensure the following inequality:

$$(1 - P_{fp})^{M_g} \geq \alpha_g, \text{ or } P_{fp} \leq (1 - \alpha_g^{\frac{1}{M_g}}). \tag{5.8}$$

Moreover, since P_{fp} is monotonically decreasing and thus $(1 - P_{fp})^{M_g}$ is monotonically increasing with respect to the number of missing tags m_g, meaning that $m_g = M_g$ makes the detection hardest and any m_g larger than M_g will ease the hardness, we thus consider the case where $m_g = M_g$ for $1 \leq g \leq G$ to meet the detection reliability regardless of m_g.

Injecting (5.8) into (5.3) with $m_g = M_g$ leads to

$$f \geq \frac{-(|\mathbb{E}| - M)k}{\ln\left[1 - \left(1 - \alpha_g^{\frac{1}{M_g}}\right)^{\frac{1}{k}}\right]},$$

where $M = \sum_{g=1}^{G} M_g$. For clarity, let $X_g \triangleq 1 - \alpha_g^{\frac{1}{M_g}}$. Because f needs to be set such that the required detection reliability for any group is achieved and k is identical for all groups, we have:

$$f = \frac{(|\mathbb{E}| - M)k}{-\ln[1 - (\min_{1 \leq g \leq G} X_g)^{\frac{1}{k}}]}. \tag{5.9}$$

Without loss of generality, let $g^* = \arg\min_g X_g$ and let the derivative of the right hand side of (5.9) with respect to k be 0, we can derive that

$$k^* = \frac{\ln \min_g X_g}{\ln \frac{1}{2}} = \frac{\ln\left(1 - \alpha_{g^*}^{\frac{1}{M_{g^*}}}\right)}{\ln \frac{1}{2}}.$$

It can be easily checked that f achieves its minimum as (5.4) at k^*. The theorem is thus proved. □

Remark Given the practical meaning of k^* and f^*, both of them should been further rounded to the smallest integers not smaller than themselves.

5.4   The Adaptive Approach

In B-detect, the reader uses the same parameters in each region, particularly the length of the Bloom filter, which may not be optimum in the case when the tags are not evenly distributed across interrogation regions. Motivated by this observation, we develop another missing tag detection protocol, named *AB-detect*, which enables the reader to use different parameters based on the number of tags in the region the reader queries.

5.4.1   Protocol Description

Phase 1: Query and Feedback Collection The reader performs a query in each region r with the parameter $(f_r, \{k_g\}_{g=1}^{G}, s)$ where f_r is the length of the Bloom filter vector used in region r, k_g is the number of hash functions used by tags in group g, s is the hash seed which is identical for all groups and regions. There are two differences compared to the baseline approach. First, f_r may be different across different regions but identical across groups; Second, k_g may be different across different groups but identical across regions. We require f_r to be a power-multiple of two, i.e., $f_r = 2^{b_r}$, $(b_r \in \mathbb{N})$. As in B-detect, the reader constructs an f_r-bit Bloom filter vector B_r with the responses from the tags in each region r. Without loss of generality, we assume that $f_1 \leq f_2 \leq \cdots \leq f_R$.

Phase 2: Virtual Bloom Filter Construction and Missing Event Detection After interrogating all R regions, the reader first expand B_r to an f_R-bit padded Bloom filter by repeating B_r $\frac{B_R}{B_r}$ times. Denote the padded Bloom filter as PB_r. The reader then combines PB_r $(1 \leq r \leq R - 1)$ and B_R to a virtual Bloom filter B^{AB} by ORing each bit of them, i.e., $B^{AB}(i) = PB_1(i) \vee \cdots \vee PB_{R-1}(i) \vee B_R(i)$ $(1 \leq i \leq f_R)$, as illustrated in Fig. 5.1. The reader then performs membership test. For each tag in group g, the reader maps its ID into k_g bits at positions $(h_v(ID) \mod f_R)$ $(1 \leq v \leq k_g)$. If all the corresponding bits in B^{AB} are 1, then the tag is regarded as present. Otherwise, the tag is considered to be missing. The reader reports a missing event for group g if the number of missing tags in the group g is at least M_g and no missing event otherwise.

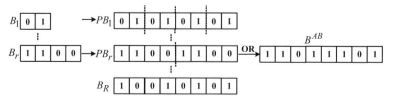

Fig. 5.1 An illustrative example of constructing virtual Bloom filter

The following lemma proves that there is no false negative in AB-detect.

Lemma 5.1 *There is no false negative in AB-detect.*

Proof It suffices to prove that if a tag is present, it holds that

$$B^{AB}(h_v(a) \bmod f_R) = 1, \ 1 \le v \le k,$$

where a denotes the ID of the tag.

Without loss of generality, assume that the tag a is located in region r. Consider any $v \le k$, let

$$h_v(a) = x + y f_r, \ x, y \in \mathbb{N}, x < f_r.$$

Let $c = \frac{f_R}{f_r}$. By definition of B_r, PB_r and B^{AB}, we have

$$B^{AB}(x + y' f_r) = PB_r(x + y' f_r) = B_r(x) = 1, \tag{5.10}$$

for $\forall y' \in \mathbb{N}, y' < c$. On the other hand, we have

$$h_v(a) \bmod f_R = x + y f_r \bmod (c f_r) = x + (y \bmod c) f_r.$$

It then follows from (5.10) that

$$B^{AB}(h_v(a) \bmod f_R) = 1.$$

The proof is thus completed. □

5.4.2 Performance Optimisation and Parameter Tuning

In this section, we investigate how to tune the parameters in AB-detect to minimise the execution time while ensuring the reliability requirement of each group. We first formulate the false positive rate for each group g, defined as $P_{fp,g}$. Recall the construction of B^{AB} in AB-detect, the probability that any bit in B^{AB} is zero is $\prod_{r=1}^{g} \left(1 - \frac{1}{f_r}\right)^{\sum_{g=1}^{G} k_g n_{gr}}$. The false positive rate for group g can then be derived as

$$P_{fp,g} = \left[1 - \prod_{r=1}^{R} \left(1 - \frac{1}{f_r}\right)^{\sum_{g=1}^{G} k_g n_{gr}}\right]^{k_g} \approx \left(1 - e^{-\sum_{r=1}^{R} \sum_{g=1}^{G} \frac{k_g n_{gr}}{f_r}}\right)^{k_g}. \tag{5.11}$$

The following theorem derives the optimal values of f_r and k_g that minimize the execution time while ensuring the group-wise reliability requirement.

Theorem 5.2 *The optimum Bloom filter vector size for the region r and the number of hash functions for the group g, denoted as f_r^* and k_g^*, that minimize the execution time while satisfying the detection reliability requirement for each group g regardless of m_g, are as follows:*

$$f_r^* = \frac{\sqrt{\sum_{g=1}^{G} k_g^* n_{gr}} \cdot \sum_{r=1}^{R} \sqrt{\sum_{g=1}^{g} k_g^* n_{gr}}}{\min_g Y_g^*}, \tag{5.12}$$

$$k_g^* = \frac{\ln(1 - \alpha_g^{\frac{1}{M_g}})}{\ln \frac{1}{2}}, \tag{5.13}$$

where $Y_g^* \triangleq -\ln[1 - (1 - \alpha_g^{\frac{1}{M_g}})^{\frac{1}{k_g^*}}]$. *The minimum execution time under the above setting, defined as T_{AB}^*, is:*

$$T_{AB}^* = \frac{1}{\min_{1 \le g \le G} Y_g^*} \left(\sum_{r=1}^{R} \sqrt{\sum_{g=1}^{G} k_g^* n_{gr}} \right)^2. \tag{5.14}$$

Proof By the same analysis as the proof of Theorem 5.1, we need to ensure the following inequality:

$$P_{dg} \ge (1 - P_{fp,g})^{M_g} \text{ or } P_{fp,g} \le (1 - \alpha_g^{\frac{1}{M_g}}). \tag{5.15}$$

Injecting (5.11) into (5.15) leads to

$$\sum_{r=1}^{R} \sum_{g=1}^{G} \frac{k_g n_{gr}}{f_r} \le -\ln[1 - (1 - \alpha_g^{\frac{1}{M_g}})^{\frac{1}{k_g}}], \ 1 \le g \le G.$$

For clarity, let $Y_g \triangleq -\ln[1 - (1 - \alpha_g^{\frac{1}{M_g}})^{\frac{1}{k_g}}]$. The above inequality is readily transformed to the following inequality:

$$\sum_{r=1}^{R} \sum_{g=1}^{G} \frac{k_g n_{gr}}{f_r} \le \min_g Y_g.$$

Without loss of generality, let $g_m = \arg \min_g Y_g$. It can be checked that

$$k_g \ge k_{g_m} \frac{\ln(1 - \alpha_g^{\frac{1}{M_g}})}{\ln(1 - \alpha_{g_m}^{\frac{1}{M_g}})}, \ 1 \le g \le G. \tag{5.16}$$

Next we derive the execution time of AB-detect, defined as T_{AB}. We can write T_{AB} as

$$T_{AB} = R \cdot G \cdot t_1' + \sum_{r=1}^{R} f_r \simeq \sum_{r=1}^{R} f_r,$$

where t_1' denotes the time for the reader to broadcast protocol parameters including the group ID for each group. In a large RFID system, it holds that $f_r \gg t_1'$. As RGt_1' is constant, finding the optimum k_g and f_r is equivalent to solving the following optimisation problem:

$$\text{Minimize: } T_{AB}' = \sum_{r=1}^{R} f_r \qquad (5.17)$$

$$\text{Subject to: } \sum_{r=1}^{R} \sum_{g=1}^{G} \frac{k_g n_{gr}}{f_r} \leq Y_{gm}. \qquad (5.18)$$

The corresponding Lagrange function can be defined as

$$\mathcal{L}(f_r, \lambda) = \sum_{r=1}^{R} f_r + \lambda \left(\sum_{r=1}^{R} \sum_{g=1}^{G} \frac{k_g n_{gr}}{f_r} - Y_{gm} \right).$$

Solving $\nabla_{f_r, \lambda} = 0$ yields the following optimum for f_r:

$$f_r^* = \frac{\sqrt{\sum_{g=1}^{G} k_g n_{gr}} \cdot \sum_{r=1}^{R} \sqrt{\sum_{g=1}^{G} k_g n_{gr}}}{Y_{gm}}.$$

T_{AB}' thus achieves its minimum with respect to f_r as below:

$$T_{AB}'^* = \frac{\sum_{r=1}^{R} \sqrt{\sum_{g=1}^{G} k_g n_{gr}} \cdot \sum_{r=1}^{R} \sqrt{\sum_{g=1}^{G} k_g n_{gr}}}{Y_{gm}}$$

$$= \frac{1}{Y_{gm}} \left(\sum_{r=1}^{R} \sqrt{\sum_{g=1}^{G} k_g n_{gr}} \right)^2.$$

It can be checked that T_{AB}' is monotonously increasing in k_g. Recall (5.16), it holds that T_{AB}' achieves its minimum as below when the equality in (5.16) holds:

$$T_{AB}'^* = \min_{k_{gm}} \frac{k_{gm} \left(\sum_{r=1}^{R} \sqrt{\sum_{g=1}^{G} \frac{\ln(1-\alpha_g^{\frac{1}{M_g}})}{\ln(1-\alpha_{gm}^{\frac{1}{M_g}})} n_{gr}} \right)^2}{Y_{gm}}. \qquad (5.19)$$

In the above equation, $\left(\sum_{r=1}^{R} \sqrt{\sum_{g=1}^{G} \frac{\ln(1-\alpha_g^{\frac{1}{M_g}})}{\ln(1-\alpha_{gm}^{\frac{1}{M_g}})} n_{gr}} \right)^2$ is a constant. Hence,

T'_{AB} is minimized when $\frac{Y_{gm}}{k_{gm}}$ is maximized. By performing straightforward algebraic

analysis, we can derive that when $k_{gm}^* = \frac{\ln(1-\alpha_{gm}^{\frac{1}{M_{gm}}})}{\ln \frac{1}{2}}$, $\frac{Y_{gm}}{k_{gm}}$ is maximized. Hence, T'_{AB}

is minimized at $k_g^* = \frac{\ln(1-\alpha_g^{\frac{1}{M_g}})}{\ln \frac{1}{2}}$ for $1 \leq g \leq G$. Injecting k_g^* into (5.19) completes

our proof.   □

Remark As k_g^* needs to be an integer and f_r a power-multiple of two, they need
to be rounded to the smallest integer and power-multiple of two not smaller than
themselves.

5.4.3   Performance Comparison: B-Detect vs. AB-Detect

Theorem 5.3 *Given the optimum parameters in both B-detect and AB-detect, the
following relationship between the minimum execution time of B-detect T_B^* and that
of AB-detect T_{AB}^* holds: $\frac{1}{R} \leq \frac{T_{AB}^*}{T_B^*} \leq 2$.*

Proof Recall (5.4), (5.5), (5.13), (5.14) and Y_g^* in Theorem 5.2, with some algebraic
operations, it can be known that $-\ln(1 - X_{g^*}^{\frac{1}{k^*}})$ in (5.4) is equal to $\min_g Y_g^*$ and
$k^* \geq k_g^*$ for $\forall g$. We then have

$$T_{AB}^* \leq \frac{k^*}{\min_g Y_g^*} \left(\sum_{r=1}^{R} \sqrt{\sum_{g=1}^{G} n_{gr}} \right)^2.$$

Let $\overline{T}_{AB}^* \triangleq \frac{k^*}{\min_g Y_g^*} \left(\sum_{r=1}^{R} \sqrt{\sum_{g=1}^{G} n_{gr}} \right)^2$ and further recall (5.1), we have

$$\frac{\overline{T}_{AB}^*}{T_B^*} = \frac{\left(\sum_{r=1}^{R} \sqrt{\sum_{g=1}^{G} n_{gr}} \right)^2}{R * \sum_{r=1}^{R} \sum_{g=1}^{G} n_{gr}}.$$

Expanding $\left(\sum_{r=1}^{R} \sqrt{\sum_{g=1}^{G} n_{gr}} \right)^2$ leads to

$$\left(\sum_{r=1}^{R} \sqrt{\sum_{g=1}^{G} n_{gr}} \right)^2 = \sum_{r=1}^{R} \sum_{g=1}^{G} n_{gr} + \sum_{i=1}^{R-1} \sum_{r=i+1}^{R} 2 \sqrt{\sum_{g=1}^{G} n_{gi} \cdot \sum_{g=1}^{G} n_{gr}}$$

$$\leq \sum_{r=1}^{R} \sum_{g=1}^{G} n_{gr} + \sum_{i=1}^{R-1} \sum_{r=i+1}^{R} \left(\sum_{g=1}^{G} n_{gi} + \sum_{g=1}^{G} n_{gr} \right)$$

$$\leq R \sum_{r=1}^{R} \sum_{g=1}^{G} n_{gr}.$$

To guarantee that f_r is power-multiple of two, we need to at most double it. It thus holds that $\frac{\overline{T}_{AB}^*}{T_B^*} \leq 2$. On the other hand, the low bound of the ratio $\frac{\overline{T}_{AB}^*}{T^*} = \frac{1}{R}$ occurs if all tags are located in only one region. It can also be noted that $T_{AB}^* = \overline{T}_{AB}^*$ when both M_g and α_g are identical across all groups. Therefore, it holds that $\frac{1}{R} \leq \frac{T_{AB}^*}{T_B^*} \leq 2$. $\qquad\qquad\square$

Theorem 5.3 leads to the following engineering implications.

- In the worst case, AB-detect doubles the execution time compared to B-detect;
- In a large asymmetric system where the number of regions R is large, AB-detect can achieve significant performance gain.

5.5   The Group-Wise Approach

In AB-detect, the reader constructs one Bloom filter that contains the response bits of tags of all groups in the interrogation region. Mixing responses from tags of different group may cause "interference" among groups and thus may increase the detection time for certain groups. Motivated by this observation, we develop a group-wise approach, termed as *GAB-detect*, in which the reader queries one group each time and constructs group-wise Bloom filters to eliminate the inter-group interference.

5.5.1   Protocol Description

Phase 1: Query and Feedback Collection   The reader performs G queries in each region r. In the g-th query ($1 \leq g \leq G$), the reader broadcasts a tetrad (g, k_g, f_{gr}, s)

where g is the group ID of group g, k_g is the number of hash functions used by group g tags, f_{gr} is the Bloom filter size used in region r for group g, s is the hash seed which is identical for all regions and groups. Again, we require f_{gr} to be a power-multiple of two. Without loss of generality, we assume that $f_{g1} \leq f_{g2} \leq \cdots \leq f_{gR}$. When receiving the query, each tag compares its group ID with g. If the tag does not belong to the group being queried, it keeps silent and waits for the next query. Otherwise, the tag selects k_g positions $(h_v(ID) \mod f_{gr})$ $(1 \leq v \leq k_g)$ in the frame of f_{gr} slots and transmits a short response at each of the k_g slots. The reader then constructs a Bloom filter for each group g and each region r, denoted by B_{gr}^{GAB}.

Phase 2: Virtual Bloom Filter Construction and Missing Event Detection After interrogating all R regions, the reader combines B_{gr}^{GAB} $(1 \leq r \leq R - 1)$ to a virtual Bloom filter B_{g*}^{GAB} for each group g by using the expansion and combination technique in AB-detect. The reader then performs membership test for each group g by using B_{g*}^{GAB}.

5.5.2 Performance Optimisation and Parameter Tuning

In this section, we investigate how to tune protocol parameters in GAB-detect to minimise the execution time while ensuring the reliability requirement of each group. We first derive the false positive rate of GAB-detect for any group g, defined as $P_{fp,g}$. Recall the construction of B_{g*}^{GAB}, the probability that any bit in B_{g*}^{GAB} is zero is $\prod_{r=1}^{R}\left(1 - \frac{1}{f_{gr}}\right)^{k_g n_{gr}}$. Hence, the false positive rate for group g can be derived as

$$P_{fp,g} = \left[1 - \prod_{r=1}^{R}\left(1 - \frac{1}{f_{gr}}\right)^{k_g n_{gr}}\right]^{k_g} \approx \left(1 - e^{-\sum_{r=1}^{R}\frac{k_g n_{gr}}{f_{gr}}}\right)^{k_g}. \tag{5.20}$$

The following theorem derives the optimal values of f_{gr} and k_g that minimize the execution time while ensuring the group-wise reliability requirement.

Theorem 5.4 *The optimum Bloom filter vector size and number of hash functions for group g in region r, denoted as f_{gr}^* and k_g^*, that minimize the execution time while satisfying the detection reliability requirement for each group g regardless of m_g, are:*

$$f_{gr}^* = \frac{\sqrt{n_{gr}} \cdot \sum_{r=1}^{R}\sqrt{n_{gr}}}{Z_g^*}, \tag{5.21}$$

$$k_g^* = \frac{\ln(1 - \alpha_g^{\frac{1}{M_g}})}{\ln \frac{1}{2}}, \tag{5.22}$$

The minimum execution time under the above setting, defined as T^*_{GAB}, is:

$$T^*_{GAB} = \sum_{g=1}^{G} \frac{\left(\sum_{r=1}^{R} \sqrt{n_{gr}}\right)^2}{Z^*_g}, \tag{5.23}$$

where $Z^*_g \triangleq \frac{\ln[1-(1-\alpha_g^{\frac{1}{M_g}})^{\frac{1}{k^*_g}}]}{-k^*_g}$.

Proof By the same analysis as the proof of Theorem 5.1, we need to ensure the following inequality:

$$P_{fp,g} \leq (1 - \alpha_g^{\frac{1}{M_g}}). \tag{5.24}$$

Injecting (5.20) into (5.24) leads to

$$\sum_{r=1}^{R} \frac{k_g n_{gr}}{f_{gr}} \leq \frac{-\ln[1 - (1 - \alpha_g^{\frac{1}{M_g}})^{\frac{1}{k_g}}]}{k_g}.$$

For clarity, let $Z_g \triangleq \frac{-\ln[1-(1-\alpha_g^{\frac{1}{M_g}})^{\frac{1}{k_g}}]}{k_g}$.

Furthermore, the execution time of GAB-detect, defined as T_{GAB}, can be derived as follows

$$T_{GAB} = RCt'_1 + \sum_{g=1}^{G}\sum_{r=1}^{R} f_{gr} \simeq \sum_{g=1}^{G}\sum_{r=1}^{R} f_{gr}.$$

Finding the optimum f_{gr} and k_g is equivalent to solving the following optimisation problem:

$$\text{Minimize: } T'_{GAB} = \sum_{g=1}^{G}\sum_{r=1}^{R} f_{gr} \tag{5.25}$$

$$\text{Subject to: } \sum_{r=1}^{R} \frac{n_{gr}}{f_{gr}} \leq Z_g, \ 1 \leq g \leq G. \tag{5.26}$$

The above optimization problem can be further decomposed to G sub-problem where sub-problem g $(1 \leq g \leq G)$ is specified as below:

$$\text{Minimize: } \sum_{r=1}^{R} f_{gr}$$

$$\text{Subject to: } \sum_{r=1}^{R} \frac{n_{gr}}{f_{gr}} \le Z_g.$$

We use the method of Lagrange multiplier to solve each sub-problem g. The Lagrange function can be defined as

$$\mathcal{L}(f_{gr}, \lambda_g) = \sum_{r=1}^{R} f_{gr} + \lambda_g \left(\sum_{r=1}^{R} \frac{n_{gr}}{f_{gr}} - Z_g \right). \tag{5.27}$$

Solving $\nabla_{f_{gr}, \lambda_g} = 0$ yields the following optimum:

$$f_{gr} = \frac{\sqrt{n_{gr}} \cdot \sum_{r=1}^{R} \sqrt{n_{gr}}}{Z_g^*},$$

where Z_g^* is the maximum of Z_g achieved at $k_g^* = \frac{\ln(1 - \alpha_g^{\frac{1}{M_g}})}{\ln \frac{1}{2}}$. Injecting k_g^* into T_{GAB} yields the optimum of T_{GAB} and completes the proof. $\qquad\square$

5.5.3   Performance Comparison: AB-Detect vs. GAB-Detect

In this section, we compare the execution time of AB-detect and GAB-detect.

Theorem 5.5 *When f_r^* in (5.12) and f_{gr}^* in (5.21) are power-multiples of two, it holds that $T_{AB}^* \ge T_{GAB}^*$.*

Proof Recall Y_g in Theorem 5.2 and Z_g^* in Theorem 5.4 and let $x_{gr} \triangleq k_g^* n_{gr}$, we can rearrange (5.23) as

$$T_{GAB}^* = \sum_{g=1}^{G} \frac{\left(\sum_{r=1}^{R} \sqrt{k_g^* n_{gr}} \right)^2}{Y_g^*} \le \frac{\sum_{g=1}^{G} \left(\sum_{r=1}^{R} \sqrt{k_g^* n_{gr}} \right)^2}{\min_g Y_g^*}$$

$$= \frac{1}{\min_g Y_g^*} \left(\sum_{r=1}^{R} \sum_{g=1}^{G} x_{gr} + 2 \sum_{i=1}^{R-1} \sum_{r=i+1}^{R} \sum_{g=1}^{G} \sqrt{x_{gi} x_{gr}} \right).$$

On the other hand, we can expand (5.14) as

$$
T_{AB}^* = \frac{1}{\min_g Y_g^*} \left(\sum_{r=1}^{R} \sum_{g=1}^{G} x_{gr} + 2 \sum_{i=1}^{R-1} \sum_{r=i+1}^{R} \sqrt{\sum_{g=1}^{G} x_{gi}} \sqrt{\sum_{g=1}^{G} x_{gr}} \right)
$$

Furthermore, define $\beta_{ir} = \sqrt{\sum_{g=1}^{G} x_{gi} \sum_{g=1}^{G} x_{gr}}$ and $\phi_{ir} = \sum_{g=1}^{G} \sqrt{x_{gi} x_{gr}}$, we have:

$$
\phi_{ir}^2 = \sum_{g=1}^{G} x_{gi} x_{gr} + \sum_{g=1}^{G-1} \sum_{w=g+1}^{G} 2\sqrt{x_{gi} x_{gr} \cdot x_{wi} x_{wr}} \tag{5.28}
$$

$$
\beta_{ir}^2 = \sum_{g=1}^{G} x_{gi} x_{gr} + \sum_{g=1}^{G-1} \sum_{w=g+1}^{G} (x_{gi} x_{wr} + x_{gr} x_{wi}), \tag{5.29}
$$

It follows from $x_{gi} x_{wr} + x_{gr} x_{wi} \geq 2\sqrt{x_{gi} x_{gr} \cdot x_{wi} x_{wr}}$ that $\phi_{ir}^2 \leq \beta_{ir}^2$. We then have

$$
(T_{AB}^* - T_{GAB}^*) \min_g Y_g^* = 2 \sum_{i=1}^{R-1} \sum_{r=i+1}^{R} (\beta_{ir} - \phi_{ir}) \geq 0.
$$

The proof is thus completed.   □

5.6   Discussion

In this section, we discuss some implementation issues of our proposed missing tag detection algorithms.

5.6.1   Estimating Tag Population

In our algorithms, the reader needs to estimate the number of tags in n_{gr} in each region and for each group. This may lead to extra overhead prior to missing tag detection. However, this overhead can be limited as the estimation can be achieved in $O(\log n_{gr})$ time using state-of-the-art estimation approaches. Specifically, we can apply two types of methods to estimate n_{gr}: single-group estimator and multi-group estimator. In the single-group estimator, when staying at region r the reader queries with the group ID g and only the tags from g respond. Then it operates like a single-group system. n_{gr} can be estimated by the methods in [5]. On the other hand, multi-group estimator estimates multiple group sizes simultaneously by employing the maximum likelihood estimation method as in [6], which is time-efficient.

Despite the extra overhead due to estimation of n_{gr}, this estimation phase enables the pre-detection of missing tags if the number of missing tags is important (e.g., due to unexpected loss or accidents). More specifically, the reader can achieve pre-detection by comparing the bitmaps constructed by the tag feedbacks and computed a priori by the reader. If a bit that is 1 in the pre-calculated bitmap by reader but turns out to be 0 in the bitmap of the feedbacks, the reader can identify the absence of tags mapped into this slot. If the number of missing tag for a given group exceeds the threshold, a missing event is reported for the group. Consequently, the reader may not need to execute the fine-grained detection algorithms as developed in the last three sections since missing tag events have already been detected in the estimation phase, thus reducing the time cost.

We may wonder whether existing tag estimation algorithms can be used to detect the missing tag event. When the detection requirement is not stringent, e.g., there are a large number of missing tags and the reader only needs to detect a small number of them so as to report a missing event, estimating the number of tags may be used. However, when the detection requirement is stringent, estimating the number of tags is not efficient as it either requires long execution time or cannot satisfy the detection requirement. To demonstrate this, we have conducted more experiments by comparing our approach with the estimation of tag numbers. Under the same detection reliability requirement, the estimation algorithm spends over 48–72.6 times as much time as our algorithms. Therefore, in our approach, we perform a coarse estimation on the tag population for two reasons: (1) our algorithms need a coarse estimation of tag population to configure parameters; (2) in case when the detection requirement is not stringent, this phase allows the reader to quickly detect a missing event.

5.6.2   Presence of Unknown/Unexpected Tags

Unknown and unexpected tags can be interpreted as the tags that have not been identified by the reader [7], such as newly arrived products, on which the reader does not have any knowledge. During the interrogation, the unknown tags will respond together with the known tags, which results in the interference to the detection of missing known tags and thus degrades the performance [8, 9].

Fortunately, two of our proposed algorithms, AB-detect and GAB-detect, are resistant to the interference caused by unknown tags. The reason is as follows. The unknown tags have not been identified by the reader, so they do not have their individual group IDs [4] such that no group ID in the interrogation messages matches with theirs. Therefore, unknown tags stay silent during the whole detection process.

5.7 Numerical Results

In this section, we evaluate the performance of the proposed approaches in terms of execution time and investigate tradeoffs under different parameter settings.

5.7.1 Simulation Settings

We conduct the experiment under both symmetric and asymmetric scenarios under different settings of R, G and M_g. By symmetric/asymmetric, we mean that tag population size in each region r is identical/different. Moreover, we set the same M_g for all group g but vary α_g for each group. Moreover, we use the symmetric transmission rate as in [10, 11] in the numerical analysis and set the transmission time for one bit to be one slot, i.e., $\delta = 1$. The length of group ID is set to $\lceil \log_2 G \rceil$ bits as in [4]. We simulate the optimum parameters settings derived in (5.4) (5.5) for B-detect, (5.12) (5.13) for AB-detect, and (5.21) (5.22) for GAB-detect.

For a comprehensive evaluation, we simulate four cases with different combination of (R, G) in both the symmetric and asymmetric scenarios: case 1: (6, 6), case 2: (12, 6), case 3: (6, 12), and case 4: (12, 12). The required detection reliability for group g ($1 \leq g \leq G$) is set to $\alpha_g = 0.749 + 0.05(g - 1)$, i.e., $0.749 \leq \alpha_g \leq 0.999$ in case 1 and case 2, and $\alpha_g = 0.44 + 0.05(g - 1)$, i.e., $0.449 \leq \alpha_g \leq 0.999$ in case 3 and case 4. The total number of tags in each region is 12,000 and the group size is $12,000/G$ in symmetric scenario. In the asymmetric scenario, on the other hand, the total number of tags is randomly chosen from [1000, 5000] in each of the first $R/2$ regions and [10,000, 20,000] in the remaining regions, and the group size in the same region is identical. The simulation results are obtained by taking the average of 100 independent trials.

5.7.2 Performance Evaluation

5.7.2.1 Performance Under Symmetric Scenario

Figure 5.2 depicts the execution time of three protocols under different threshold for the four cases in the symmetric scenario. As shown in the results, globally GAB-detect achieves the best time efficiency and AB-detect outperforms B-detect, especially when the detection reliability for each group varies more significantly, i.e., $G = 12$. This can be explained as follows: The frame size in AB-detect and B-detect are set based on $\min_g Y_g^*$ in Theorem 5.1 and 5.2, which overkills the groups with larger Y_g^*. In contrast, GAB-detect addresses this limit by eliminating the inter-group interference. We can also observe that in some cases, GAB-detect has longer execution time than AB-detect. This is due to the design requirement that the Bloom filter size needs to be the power-multiple of two. However, globally speaking, GAB-

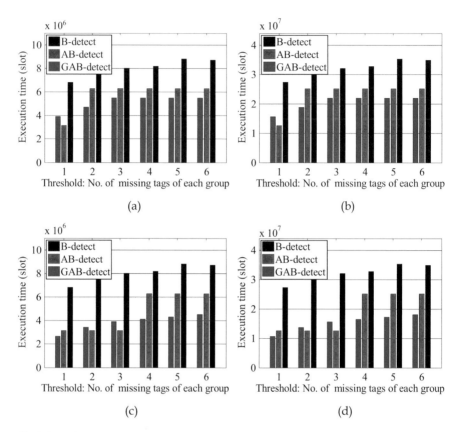

Fig. 5.2 Performance comparison in symmetric scenario. (**a**) Case 1: $R = 6$, $G = 6$, (**b**) Case 2: $R = 12$, $G = 6$, (**c**) Case 3: $R = 6$, $G = 12$, (**d**) Case 4: $R = 12$, $G = 12$

detect still outperforms B-detect. Furthermore, we investigate the actual reliability of the proposed schemes. The results demonstrate that all proposed schemes can detect the missing event with probability one.

5.7.2.2   Performance Under Asymmetric Scenario

Figure 5.3 illustrates the execution time for the four cases with different thresholds in the asymmetric scenario. It can be seen from the four subfigures in Fig. 5.3 that GAB-detect outperforms AB-detect and saves execution time up to 70% in comparison to B-detect. This can be interpreted as follows: In the asymmetric scenarios, the performance gap between AB-detect and B-detect is more significant compared to the symmetric scenario because the frame size in B-detect is identical across the regions regardless of the tag size in an individual region while AB-detect distinguishes the regions with different tag sizes when setting the frame size. Furthermore,

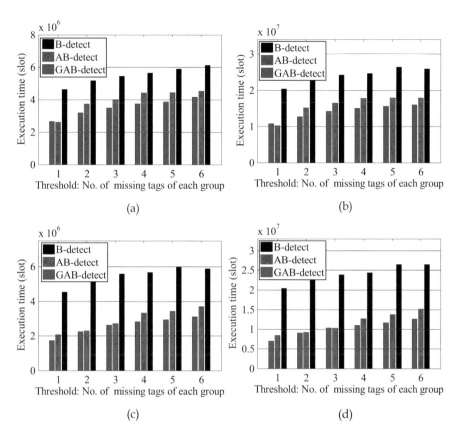

Fig. 5.3 Performance comparison in asymmetric scenario. (**a**) Case 1: $R = 6$, $G = 6$, (**b**) Case 2: $R = 12$, $G = 6$, (**c**) Case 3: $R = 6$, $G = 12$, (**d**) Case 4: $R = 12$, $G = 12$

we investigate the actual reliability of the proposed schemes and the results show that all proposed schemes can detect the missing event with probability one.

To further evaluate the performance and evaluate the analytical results, we conduct a set of numerical analysis in a even more asymmetric scenario where the tag size is randomly chosen from [50, 100] in each of the first $R − 1$ regions and from [5000, 10,000] in the remaining region. As shown in the four subfigures in Fig. 5.4, the performance gain of GAB-detect and AB-detect over B-detect is more remarkable. Specifically, the detection time of B-detect is up to 12.6 times as much as that of GAB-detect and AB-detect.

5.7.2.3 Impact of Nonidentical M_g

To comprehensively evaluate the performance, we conduct more numerical analysis in both symmetric and asymmetric scenarios which are same with the previous

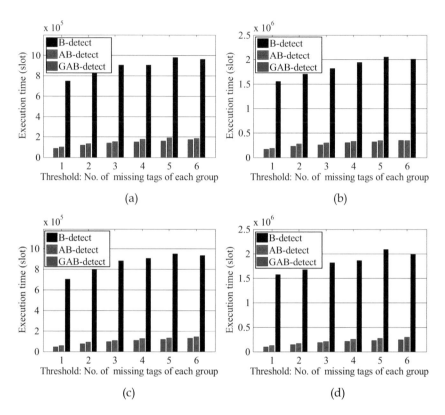

Fig. 5.4 Performance comparison in more asymmetric scenario. (**a**) Case 1: $R = 6$, $G = 6$, (**b**) Case 2: $R = 12$, $G = 6$, (**c**) Case 3: $R = 6$, $G = 12$, (**d**) Case 4: $R = 12$, $G = 12$

Table 5.2 Execution time ($\times 10^6$) under nonidentical M_g and ϵ

	Number of groups G		
Scenario	6	12	Estimation error ϵ
Symmetric	(8.1, 6.3, 4.7)	(8.4, 6.3, 4.3)	0
	(8.9, 6.3, 4.7)	(9.3, 6.3, 4.6)	10%
Asymmetric	(6.1, 3.9, 3.2)	(6.4, 3.1, 2.8)	0
	(6.6, 4.2, 3.4)	(6.9, 3.4, 3.1)	10%

settings except that R is fixed to 6 and $M_g = g$ for group g. Moreover, we also investigate the impact of estimation error ϵ on the performance.

From the results listed in Table 5.2, we can observe that GAB-detect significantly outperforms AB-detect and B-detect when M_g is different for each group. Besides, the execution time increases by less than 11% in the worst case when ϵ varies from 0 to 10%. While on average, B-detect and GAB-detect and AB-detect spend 9% and 6% and 2.6% more time, respectively. Therefore, it is fair to allow $\epsilon = 10\%$.

5.8   Conclusion

In this chapter, we have formulated a missing tag detection problem arising in multiple-group multiple-region RFID systems, where a mobile reader needs to detect whether there is any missing event for each group of tags. By leveraging the technique of Bloom filter, we develop a suite of three missing tag detection protocols, each decreasing the execution time compared to its predecessor by incorporating an improved version of the Bloom filter design and parameter tuning. In our future work, we plan to study the case where multiple mobile readers are available to detect missing tag events and design optimum missing tag detection algorithms in that context.

References

1. H. Liu, W. Gong, X. Miao, K. Liu, W. He, Towards adaptive continuous scanning in large-scale rfid systems, in *IEEE INFOCOM* (IEEE, Piscataway, 2014), pp. 486–494
2. L. Xie, Q. Li, C. Wang, X. Chen, S. Lu, Exploring the gap between ideal and reality: an experimental study on continuous scanning with mobile reader in RFID systems. IEEE Trans. Mob. Comput. **14**(11), 2272–2285 (2015)
3. B.H. Bloom, Space/time trade-offs in hash coding with allowable errors. Commun. ACM **13**(7), 422–426 (1970)
4. J. Liu, B. Xiao, S. Chen, F. Zhu, L. Chen, Fast rfid grouping protocols, in *IEEE IFOCOM* (2015), pp. 1948–1956
5. B. Chen, Z. Zhou, H. Yu, Understanding RFID counting protocols, in *ACM MobiHoc* (2013), pp. 291–302
6. W. Luo, Y. Qiao, S. Chen, An efficient protocol for rfid multigroup threshold-based classification, in *IEEE INFOCOM* (2013), pp. 890–898
7. X. Liu, B. Xiao, S. Zhang, K. Bu, Unknown tag identification in large RFID systems: an efficient and complete solution. IEEE Trans. Parallel Distrib. Syst. **26**(6), 1775–1788 (2015)
8. M. Shahzad, A.X. Liu, Expecting the unexpected: fast and reliable detection of missing RFID tags in the wild, in *IEEE INFOCOM* (2015), pp. 1939–1947
9. J. Yu, L. Chen, K. Wang, Finding needles in a haystack: Missing tag detection in large RFID systems (2015, Preprint). arXiv :1512.05228
10. M. Chen, W. Luo, Z. Mo, S. Chen, Y. Fang, An efficient tag search protocol in large-scale rfid systems, in *IEEE INFOCOM* (2013), pp. 899–907
11. W. Luo, S. Chen, Y. Qiao, T. Li, Missing-tag detection and energy–time tradeoff in large-scale RFID systems with unreliable channels. IEEE/ACM Trans. Netw. **22**(4), 1079–1091 (2014)

Chapter 6
Conclusion and Perspective

6.1 Book Summary

Recent years have witnessed an unprecedented development of the radio frequency identification (RFID) technology due to its low cost and non-line-of-sight communication pattern. RFID tags are becoming ubiquitously present in retail products, library books, debit cards, passports, driver licenses, car plates, medical devices, etc. As a result, RFID technology is shaping up to be an important building block for the Internet of Things (IoT). In most, if not all, RFID applications, tag management and monitoring are perhaps one of the most fundamental components. Although simple to state and intuitively understandable, designing efficient tag management and monitoring algorithms require non-trivial efforts to achieve, especially in large-scale RFID systems, due to the particular challenges posed by the RFID paradigm such as the system dimension in terms of tag population and dynamics, the limited computing resource at individual tags, the notoriously unreliable wireless links, and the invasive and ubiquitous nature of tags.

The above constraints and considerations make tag management and monitoring in this context an emerging research field requiring new tools and methodologies. In this regard, by the present book we hope to make a tiny while systematic step forwards in the design and analysis of tag management and monitoring algorithms that can scale elegantly, act efficiently in terms of time and energy.

Specifically, this book has been dedicated to addressing the fundamental problem of tag counting and monitoring in large-scale RFID systems, at both the theoretical modeling and analysis and the algorithm design and optimisation levels, with Chap. 2 focusing on the stability analysis of FSA protocol, Chap. 3 proposing a tag population estimation framework in dynamic RFID systems, Chap. 4 addressing miss tag event detection problem in the presence of unexpected tags, and Chap. 5 devising a suit of algorithms for multiple-group multiple-region RFID systems to detect missing tag event. More specifically, Chap. 2 presented complete and accurate characterisation of FSA behavior, which provides theoretical guidelines

J. Yu, L. Chen, *Tag Counting and Monitoring in Large-Scale RFID Systems*, https://doi.org/10.1007/978-3-319-91992-8_6

on the design of stable FSA-based protocols in other practical applications such as vehicular networks and M2M networks. In Chap. 3, we tackled the dynamic tag population estimation problem and showed a theoretical method of analyzing its estimation error and convergence rate. Furthermore, we illustrated the key to algorithm design and parameter configuration in missing tag event detection problem in Chaps. 4 and 5.

In what follows, we discuss a number of open questions we judge pertinent to the topics addressed in the book and outline several important potential directions for future research.

6.2 Open Questions and Future Work

6.2.1 Energy Efficiency

The state-of-the-art research on RFID mainly focuses on improving time efficiency. This metric is suited for short-range passive tags. However, in more generic context such as large-scale infrastructure systems, an RFID system often covers a large area and involves a large number of tags. In this case, battery-powered active tags are preferred for longer transmission ranges and richer on-tag hardware for more sophisticated functions. Hence, energy efficiency becomes a primary concern because it determines the tags' lifetime before they have to be recharged. A critical research direction is to design energy-time-efficient algorithms for such large-scale tagged systems and investigate new methods for controlling the tradeoff between the algorithm execution time and the energy cost.

A concrete example in this regard is the problem of information collection, a fundamental functionality allowing readers to collect control information and data from tags so as to manage and monitor the whole RFID network. Information collection is also one of the most energy-consuming operations as it involves transmission of data form either battery-powered tags or passive tags energized by the radio wave of the readers. Therefore, energy efficiency is a primary concern. A potential direction is to exploit the data redundancy to improve the energy efficiency in information collection. More formally, this consists of exploring the correlation of data produced by tags close to each other and select a subset of tags each time to collect data so as to reduce the number of responding tags. Collecting data from only a subset of tags also improves the time efficiency of the information collection. However, the price for the performance gain is the accuracy of the collected data. Therefore, we need to quantify this trade-off between energy efficiency and the error of the collected data, based on which we develop information collection algorithms optimizing energy efficiency while bounding expected error.

Another research axis is to leverage wireless charging technology to prolong tags' lifetime. This technology allows to remotely recharge batteries, but induces new costs inherent to this technology. Investigating this area allows to propose solution to complex optimization problems to reach the best cost-efficiency trade-offs: How to determine the number of mobile wireless chargers with respect to the area of coverage and cost? What is the trajectory of those mobile chargers? What kind of correlations exist between recharging and data collection and how to leverage that to reach the best delay-cost trade-offs?

6.2.2 Security and Privacy

With pervasively deployed tags, privacy becomes a critical concern. While RFID tagged systems provide a powerful platform for collecting statistics, we want to do so without tracking each individual tag or causing other privacy violation. Given limited capacity of tags on both processing and communication, designing secure and privacy-preserving tag management algorithms is by no means trivial and often needs to balance between several design metrics such as security, performance, scalability and energy efficiency.

6.2.3 Compatibility and Implementability

As the third perspective, we would like to put a particular emphasis on the compatibility and the implementability of the developed algorithms. This is motivated by the observation that many tag management and monitoring algorithms proposed in the literature are based on elegant theoretical tools but fail to be compatible with the current and upcoming RFID standards, thus significantly limiting their implementability in practice. Therefore, it is important to take the compatibility as an explicit design metric such that the developed algorithms can be directly implemented (or after lightweight modifications) in the off-the-shelf commodity tags.

Methodologically, the study on RFID systems is naturally multidisciplinary and calls for tools in algorithm design and analysis, security and privacy, wireless channel modeling, system and hardware development and test. We expect to see more in-depth works that mobilize a large spectrum of tools to design practical and implementable algorithms in the emerging RFID systems.

Index

© Springer International Publishing AG, part of Springer Nature 2019   133
J. Yu, L. Chen, *Tag Counting and Monitoring in Large-Scale RFID Systems*,
https://doi.org/10.1007/978-3-319-91992-8

Printed in the United States
By Bookmasters